Monitoring and Compliance:
The Political Economy
of Inspection

POLITICAL ECONOMY AND PUBLIC POLICY,
VOLUME 4

Editors: William Breit, *Department of Economics, Trinity University, San Antonio*
Kenneth G. Elzinga, *Department of Economics, University of Virginia*

POLITICAL ECONOMY AND PUBLIC POLICY

An International Series of Monographs
in Law and Economics, History of Economic Thought
and Public Finance

Edited by William Breit, Department of Economics, Trinity University and
Kenneth G. Elzinga, Department of Economics, University of Virginia

For Marge and Henry
Who, for many years, monitored my behavior

Monitoring and Compliance:
The Political Economy
of Inspection

by: DAVID HEMENWAY
*Department of Health Policy
and Management
Harvard School of Public Health*

 JAI PRESS INC.

Greenwich, Connecticut   *London, England*

Library of Congress Cataloging in Publication Data

Hemenway, David.
    Monitoring and compliance.

    (Political economy and public policy ; v. 4)
    Bibliography: p.
    Includes index.
    1. Labor inspection—United States.   2. Industrial
    safety—United States.   3. Industrial hygiene—United
    States.   4. Environmental monitoring—United States.
    I. Title.   II. Series.
HD3665.H46   1985   353.0082   84-29980
ISBN 0-89232–477–5

CONTENTS

Preface

This book is about mandatory inspection. It compares and contrasts a large number of inspection processes from the viewpoint of the individual inspector. The aim of the book is to provide useful insights and to stimulate thought about this important aspect of regulation.

The book presents a conceptual framework for thinking about inspectors and the inspection process. The emphasis is on how the inspection environment affects the inspector's behavior and performance. Many examples are provided to illustrate the suggested relationships. These examples are often historical and do not necessarily describe current regulations or inspection procedures, for both laws and policies change.

This is a social science, not a policy document. The aim is not to evaluate agency performance nor to instruct managers on effective agency administration. The book does not explain how to do a good inspection, nor does it advise inspectors on proper behavior. Instead the purpose is to try to explain why inspectors seem to do a better job in some circumstances than in others.

The book is intended for social scientists—economists, political scientists, and sociologists. Legislators, regulators, inspectors and policy analysts should find it informative. It is also hoped that others might enjoy it and learn something from it. An attempt has been made to make the book short, readable and interesting. Jargon has been kept to a minimum.

The book always uses the masculine pronoun. This is done for stylistic reasons. I apologize for any offense.

I would like to thank the numerous inspectors who talked to us, and allowed us to accompany them on the job.

Many people read all or portions of the manuscript and made valuable comments: Alexia Antczak, Les Boden, Graham Colditz, Tish Davis, Penny Feldman, Margaret Gerteis, Paula Griswold, Cynthia King, Trevor King, Marc Roberts, and Steve Thomas.

I had three excellent research assistants, Suzanne Cashman, Beth O'Brien and Rob Lunbeck. Rob was on the project the longest, went on many inspections, and provided numerous insights and ideas.

The research was financed by a grant from the Henry J. Kaiser Family Foundation, Menlo Park, California. I am very grateful for their generous support.

David Hemenway
Harvard University

Chapter I

Introduction

Regulation is a governmental service whose 'technology' of production
is disarmingly simple. Some person or persons in a position of authority
issue a prescriptive command to a designated or identifiable source.
Someone then examines the behavior of the members of that audience
to ascertain whether they are complying with the command. In the
event that a member is found not to have been in compliance, someone
may impose a painful sanction upon him.[1]

Of the three parts of regulation—prescription, inspection and
enforcement—the first and third have received substantial analysis. Minimum standards versus incentive pricing, for example, has been an important theme in the academic literature.[2]
The appropriate level and gradation of enforcement penalties
and sanctions have also been studied.[3] By contrast, inspection
as a separate entity in the regulatory process has received little
attention.[4] This book is intended to help remedy that situation.

The inspection link is a crucial one, a link that too often is
neglected by policy analysts. The debate over nuclear safety,
for example, took on a new dimension when the issue tran-
scended the theoretical question of fail-safe systems and consideration was given to the practical problems of inspection.
Whether nuclear power plants can be made safe is now recognized as an issue for social as well as physical scientists.

Inspections provide the information by which we judge the
level of compliance with regulations and hence their efficacy.
This in turn influences policy prescriptions. But inspection re-

1

ports are not always accurate. Often they are biased in particular directions. An understanding of the inspection process can help us to recognize and perhaps even eliminate some of these biases.

In the early 1970s, the Department of Interior reported that 94% of the active underground coal mine sections in the United States were meeting the 2.0 milligram standard established by Congress as the acceptable dust level. This reassuring and precise assessment was based, naturally enough, on the Department's inspection program that annually took some half-million samples from the 3,700 active mines. The problem was that the results were largely unreliable. The equipment used did not provide accurate measurements, and correct sampling techniques were not being employed. In part, improper procedures were followed because the samples were selected and controlled by the mine operators—whose incentives were not necessarily to ensure accuracy. Moreover, the miners who took the original samples were ill-trained. Many did not understand the purpose of the program, and, further, no one enjoyed wearing the cumbersome monitoring equipment for the entire day. Some miners admitted trying to get low readings expecting that they would not have to wear the sampler as often. A few others said they tried to get high readings in the belief this would help qualify them for black lung benefits.[5] (The report on the deficiencies of the coal mine inspection program was itself based on a type of inspection—in this case, a monitoring of inspectors. There is no reason, of course, to blithely assume that that inspection report was 100% reliable.)

Inspections are ubiquitous in our society. Wherever there are regulations, there are also inspections. There are housing inspections, motor vehicle inspections, and building inspections. There are Occupational Safety and Health Administration (OSHA) inspections, Environmental Protection Agency (EPA) inspections, Consumer Product Safety Commission (CPSC) inspections, immigration and customs inspections. There are inspections of railroads, airplanes, elevators, restaurants, meat, grain, fruits and vegetables, nursing homes and power plants. There are weights and measures inspections and Professional Standards Review Organization (PSRO) monitoring. Internal Revenue Service (IRS) agents and police officers have inspection functions. There are also many private inspectors, such as the National Board of Boiler and Pressure Vessel inspectors, Underwriters' Laboratories testers, home inspectors and insurance adjusters.

This book is about public mandatory inspection. It examines a large number of inspection processes that are part of a government regulatory system. The broad purpose is to try to understand more about inspection generally. The more narrow goal is to help clarify why some inspections work better than others. This is important information for the decision about what form of regulation to adopt, and indeed, whether it makes sense to regulate at all.

The issue examined in this book is under what conditions the inspector, and his colleagues, have the ability and the incentive to do a good job. Our thesis is that the inspection environment is a key factor in determining the inspector's overall performance. The emphasis throughout the book is thus on the influence of the inspection environment (e.g., what is being inspected, the number and geographic distribution of inspectees) rather than on the effects of specific managerial practices (e.g., what are the hiring practices of the agency, the training provided to inspectors, the promotion policies). The inspection environment is more enduring, and, we believe a good understanding of it permits reasonable predictions about the expected quality of the inspection process. The book describes the most important environmental aspects and provides a framework for viewing and, we hope, understanding the advantages and drawbacks of broad regulatory approaches.

The focus of this book is the inspector rather than the agency management. Inspectors in different agencies engage in somewhat disparate activities. For example, some read written reports, some examine machines, others directly observe individual behavior. Some inspectors are aided by complicated equipment, others rely primarily on their own senses. Some inspectors face danger, others may toil in comparative safety. Yet there are certain broad similarities in the work of most inspectors. The typical government inspector, for example, is a "street level bureaucrat."[6] He meets face-to-face with the regulatee. His job normally entails the most instrusive aspect of the regulation. For many, he represents the only human contact with the government bureaucracy, and he comes to personify the agency.

A typical traveler does not experience "United States Customs Policy"; instead he encounters particular customs inspectors who examine his packages and belongings.

In some sense the inspectors do the most critical work in the agency; without them, there is no effective regulation. Some years ago, the U.S. Customs Service argued vehemently that substantial budget cuts would cripple its operation. When the appropriations were nonetheless reduced, it responded effectively by threatening to fire every customs inspector in the United States, but not one other member of the agency.[7]

Although an inspector's work is crucial, he generally occupies the lowest professional position in the regulatory agency. Compared to most low level employees, however, the inspector is largely free from effective supervision. He thus has a great deal of discretion. A police officer, for example, can have wide latitude. A decision not to cite a violation will rarely be known about, let alone questioned.[8] A judge described the awesome responsibility of immigration inspectors: "Every day, [they,] including very junior people, are required to make decisions that can affect the freedom of movement, economic condition, liberty and perhaps even the future survival" of those who enter their domain.[9]

A fundamental aspect of every inspector's job is its adversarial nature. This feature differentiates inspection from most other occupations. The inspector's work entails examining individual inspectees, who generally prefer that their specific shortcomings neither be discovered nor reported. There is some notion of a "hide-and-seek" game here, but with the added possibility of the inspector's incentives to do a good job being distorted by the regulated party. This interplay between inspector and inspectee is a primary focus of the book.

NOTES AND REFERENCES

1. Colin Diver, "A Theory of Regulatory Enforcement," *Public Policy*, Summer 1980, 28 (3), p. 4.

2. Allen V. Kneese and Blair T. Bower, "Standards, Charges and Equity," *Managing Water Quality: Economics, Technology, Institutions* (Baltimore, Johns Hopkins Press, 1968).

William J. Baumol and Wallace E. Oates, "The Use of Standards and Prices to Protect the Environment," *Swedish Journal of Economics*, 73 (March 1971), 42-54.

Marshall I. Goldman, "Pollution: The Mess Around Us," *Ecology and Economics* (Englewood Cliffs, NJ: Prentice Hall, 1972) 3-63.

Myrick A. Freeman, "The Market System and Pollution," *The Economics of Environmental Policy* (New York, NY: John Wiley & Sons, 1973) 1-16.

3. Gary Becker, "Crime and Punishment: An Economic Approach," *Journal of Political Economy* 76, No. 2 (March-April 1968) 169-217.

George Stigler, "The Optimum Enforcement of Laws," *Journal of Political Economy* 78, No. 3 (May-June 1970) 526-536.

William M. Landes, "An Economic Analysis of the Courts," *The Journal of Law and Economics*, 14, No. 1 (April 1971) 61-107.

John Ferry and Marjorie Kravitz, *Issues in Sentencing*. Washington, DC: U.S. Department of Justice, Law Enforcement Assistance Administration, March 1978.

Roland McKean, "Enforcement Costs in Environmental and Safety Regulation," *Policy Analysis*, Summer 1980, Vol. 6, No. 3.

4. An important recent exception is Eugene Bardach and Robert A. Kagan, *Going by the Book: The Problem of Regulatory Unreasonableness*. Philadelphia: Temple University Press, 1982.

5. General Accounting Office, "Improvements Still Needed in Coal Mine Dust-Sampling Program and Penalty Assessments and Collections," December 1975.

6. Michael Lipsky, *Street Level Bureaucracy* (New York, NY: Russell Sage Foundation, 1980).

7. Tom Alexander, "Why Bureaucracy Keeps Growing," *Fortune*, May 7, 1979, p. 168.

8. Egon Bittner, *Functions of the Police in Modern Society; a Review of Background Factors, Current Practices and Possible Role Models*, National Institute of Mental Health, Center for Studies of Crime and Delinquency, Chevy Chase, Maryland, 1970.

9. Federal Immigration Judge Theodore P. Jakaboski, quoted in John Crewdson, "U.S. Immigration Service Hampered by Corruption," *The New York Times*, January 13, 1980, p. 46.

Chapter II

The Inspector's Ability

Illicit traffic in rare animals is further imperilling already endangered species. One major problem is that customs officials are usually ill-equipped to spot the smuggled goods among the more than 90 million wildlife items that are annually brought into the United States. It is not easy to recognize the vial of illegal turtle oil concealed in the cosmetic case.

Newsweek, March 26, 1979

In his endeavor to help the regulatory agency accomplish its mission, the inspector may assume a number of roles—Reporter, Enforcer and Consultant. Every inspector makes examinations and presents his findings. This investigative, or what we call Reporting aspect of inspection is his primary role. Some inspectors also have direct enforcement responsibilities. Others do not. Food and Drug Administration (FDA) and EPA inspectors, for example, have no enforcement powers. Meat/poultry and elevator inspectors, on the other hand, have the authority to shut down operations.[2]

Some inspectors are also Consultants. Part of their job is to counsel and educate. Nursing home surveyors, for example, see themselves as providing advice and motivation to providers, as actually "selling health care."[2] IRS agents, by contrast, give their version of the law, but rarely help taxpayers decrease their tax liabilities.[3] Inspectors' view of themselves as consultants can

7

be somewhat self-serving. Pest-control housing inspectors in Philadelphia, for example, depicted themselves as "helping" tenants, particularly the elderly, by being friendly listeners and aiding with personal problems. The tenants, however, did not regard them in this light, instead complaining of "unannounced Gestapo-like inspections."[4]

Inspectors have some expertise, of course, and could often provide useful advice. Where their relationship to inspectees is primarily consultative rather than adversarial, however, is generally where there is little punitive enforcement. Accreditation and evaluation teams, for example, and inspectors involved in the selling of insurance to major buyers, largely give advice and guidance.[5] In the area of government radiologic health regulation of medical X-ray machines, where enforcement actions are notoriously weak, the inspector/inspectee relationship is largely consultative.[6]

This book only touches on the consultative and enforcement aspects of inspection. One major concern is with the inspector as Reporter. Our primary interest is in the inspector's ability and incentive to do a good Reporting job. It should be emphasized that making a good report is not identical to helping correct deficiencies. In housing, as we shall argue, the inspector could be an ideal Reporter, yet often find little or no improvement for his efforts. By contrast, a good elevator inspection generally ensures passenger safety.

I. INNATE ABILITY AND TRAINING OF INSPECTORS

Properly trained inspectors are essential for a successful inspection program. Quantity is rarely a sufficient substitute for quality. By 1898, Prussia had created an elaborate bureaucracy to protect its citizens from the bane of trichinosis. Some 26,000 inspectors (more than the entire enlisted ranks of the U.S. Army on the eve of the Spanish-American War) used microscopy to inspect all pork products. Expenditures for this vast undertaking exceeded the appropriations for the entire U.S. Department of Agriculture. Yet Germany still had many more cases of trichinosis than the United States, principally due to deficiencies in their inspection system. The major problem was that most of the large army of Prussian inspectors were part-time and poorly trained workers—often barbers or tradesmen, and sometimes the butchers themselves.[7]

An effective inspector typically needs some innate intelligence, specific training and a good personality. Level of education is often used as a screening device for intelligence. Requirements run the gamut from high school and less (e.g., Philadelphia pest-control inspectors, U.S. meat and poultry inspectors) to college degrees and Ph.D.'s (CPSC and Energy Research and Development Administration (ERDA) weapons production inspectors). Entering salary grades for federal inspectors range at least from GS-2 (FDA consumer safety inspectors) to GS-12 (ERDA).[8]

Specific credentials are sometimes a prerequisite for becoming an inspector. Massachusetts nursing home surveyors are all Registered Nurses with a minimum number of years of supervisory experience. USDA health inspectors of animal research facilities are all veterinarians. FAA operations inspectors must hold appropriate pilot or instructor certificates. Federal coal mine (Mining Enforcement and Safety Administration) inspectors need five years of underground mine experience.[9]

For many inspectors, "personality is the whole ball game."[10] Since an inspector usually has numerous face-to-face encounters, he needs good interpersonal skills. Not only should he be honest and just, but given the adversarial nature of his job, he often must be firm and have a thick skin. Keeping an adversarial encounter from becoming antagonistic decreases resentment against the regulations and increases the inspector's effectiveness.

Desirable personality characteristics are difficult to assess during recruitment, and since they cannot be measured objectively, they are rarely listed as formal criteria for the hiring of inspectors. One management method occasionally employed as a screening device is active on-the-job appraisal during a probationary period.

The initial training given to inspectors depends partly on the nature of the job and partly on their entering qualifications. If the agency does not provide introductory training, it must rely on the inspectors' previous education and experience. The Canadian Atomic Energy Control Board, for example, has no formal training program, but requires its personnel to have at least an M.A. degree and past nuclear experience.[11]

Many agencies provide some introductory and on-the-job training. Federal meat and poultry inspectors have a nine-month training period during which they both attend an agency school at Fort Worth, and work under the supervision of either a veterinarian or an experienced inspector. This training qualifies inspectors for work in slaughterhouses, which is the starting point for all new inspection personnel.[12]

Formal training beyond the introductory level is sometimes available. Such programs are generally linked to career development opportunities. The feasibility of career advancement affects the incentives of the inspector and is discussed in Chapter III.

II. DIFFICULTY OF THE TASK

Sufficient quantities of labor and capital are necessary to accomplish any specific task. A perennial complaint of many regulatory agencies is that they lack adequate appropriations and personnel to achieve their prescribed goals. For some inspections, the magnitude of the legislative charge makes it virtually impossible to have enough inspectors to catch a high percentage of violations. OSHA standards, for example, apply to more than five million workplaces. Yet the agency inspects only some 90,000 per year.[13] Similarly, the EPA, CPSC and IRS can scrutinize only a small fraction of their large regulated universes. By contrast, all nursing homes and every commercial airplane can be regularly and carefully inspected. (Certain agencies, such as the CPSC and the Bureau of Motor Carrier Safety, have additional inspection difficulties because firms legally under their jurisdiction need not register with them. They thus typically lack a complete list of the many firms under their authority.)[14]

The ability of the agency to do the requisite monitoring to fulfill its mandate depends not only on the sheer number of inspectees, but also on the time necessary to perform inspections. Thus, despite the tens of millions of cars in the U.S., satisfactory motor vehicle safety inspections can be done quickly, and practically every automobile could be inspected. Conversely, even though there are few nuclear plants, the enormous time and expense necessary to insure achievement of the expected high safety levels make it financially burdensome to have much more than a nominal federal inspection force.

Time constraints generally limit the ability of inspectors to discover many of the below-standard products or processes. Even if every item is examined the inspector may be too hurried to do a thorough job. In the 1960s, for example, meat inspectors were given only two seconds per carcass to determine if the meat were wholesome and free from infection.[15]

The inspector's effectiveness may also be decreased if little of his time is devoted to actual examinations. OSHA inspectors spend only about one-third of their time in the field. The rest is used to prepare for travel to, and report on their site visits.[16] Regional inspectors for the Nuclear Regulatory Commission spend only about 25% of a typical workweek at the power plant.[17] In order to increase the time available for directly observing nuclear activities, the NRC is planning to quadruple its force of *resident* inspectors. However, even for these employees, much valuable time is tied up doing clerical and other mundane work. Some resident inspectors, for instance, consume 15% of their time picking up official mail. The inspector is supposed to collect his mail daily. But important mail cannot be delivered to the utility site because the NRC fears tampering that would compromise confidential documents. The inspector's mail is therefore routed to the nearest post office—which may be 25 miles away.[18]

The ability of inspectors to detect most violations depends not only on their numbers, quality and training, but also on the capital equipment available for their use. Coal mine inspectors cannot effectively monitor air quality if their samplers do not produce reliable measures. In the mid-1970s, the National Bureau of Standards reported that the equipment, even when meticulously operated by trained scientists, yielded dust level measurements that varied as much as 32% from the actual dust concentrations in the underground mines.[19] FAA inspectors have difficulty detecting metal fatigue cracks if these defects do not show up on routine X-rays.[20] Harvard library guards will never be able to catch more than a small percentage of book thieves until a sophisticated security system is installed.[21]

The ability to monitor inspectee performance may vary over time, as the technology and the regulated universe change. Sewer inspections improved, and became easier to do, when advances in camera technology enabled every foot of sewer line to be scrutinized without digging up the sewer.[22] Chromosome tests now make it virtually impossible for a man to pass himself off as a woman at the Olympic games.[23] Income tax returns are becoming simpler to examine as more and more data arrive on computer tapes. While all the information on the tapes enters the initial screening, the IRS does not have sufficient resources to catalogue more than 15% of the data that arrive in paper form.[24] The introduction of the supertankers decreased transportation costs, but also lessened the effectiveness of the Coast Guard's normal inspection methods. To treat the huge tankers in the same manner as older, smaller ships would result in unsatisfactory examinations.[25]

The number and geographic location of inspectees are factors affecting the cost of performing inspections. The historic decline of the small private slaughterhouse, and its replacement by larger, more centralized municipal facilities, increased the inspection agency's ability to ensure a safe and wholesome product. Similarly, allocating particular times for slaughtering, and designing plants specifically to minimize the problems of monitoring, can dramatically decrease the resources required for satisfactory inspections.[26]

Some agencies have the ability—the wherewithal—to do what may be considered a satisfactory job. Others, such as the Immigration and Naturalization Service (INS), do not. The principal INS mission is to keep the U.S. free from illegal immigrants. It is responsible for patrolling the border, and for deporting aliens illegally residing in this country. A recent GAO study concluded that the INS currently has "neither the legal means nor sufficient resources to stem the growing number of illegal aliens" in the U.S.[27] It added that the future prospects for effectively enforcing the immigration laws also appear dim. A 50% increase in total agency appropriations, for example, might reasonably be directed entirely to enhancing the security of only 2.5% of the United States' 8,000-mile border. And the INS is better able to prevent unlawful entry than to find and deport the estimated five million illegal aliens living in the States. Currently, for instance, the INS does not have up-to-date reliable information on perhaps even a majority of legal aliens—such as students—in the U.S. The task of apprehending those who decide to remain here illegally is vastly more formidable.[28]

III. BALANCE OF POWER

The most significant characteristic that distinguishes inspection from most other work is its adversarial nature. Normally, the inspector's principal task is to discover violations of the agency's prescriptions. The inspectee generally prefers that few infractions be either found or reported.

This is not to imply that the inspector and inspectee do not frequently have similar interests. Both, for example, want to see that certain problems are corrected. No one wants a nuclear catastrophe, a plane crash or food poisoning. But the inspectee may want to devote less resources to preventing these problems than society deems justified. This is the main economic justification for much of the regulation in the area of health and safety.

The regulations are intended to change the firms' incentives, to induce them to act more in society's interest—to make their restaurants cleaner, their planes and power plants safer. But the firms have other possible responses to regulation, one of which is to attempt to prevent the infraction, or its severity, from being reported. The firm may try to "capture" the inspector, seducing him to act in a manner favorable to the firm, and somewhat contrary to the agency's purported mission. Or the firm may try to hide its violations. Conditions may be altered so that there are no violations at the moment of inspection, or deficiencies may not be corrected, but merely camouflaged.

The ability to hide violations (or the cost of hiding them) is just one factor that influences the inspectee's decision concerning what actions, if any, to take to avoid detection. Other factors include the costs if infractions are discovered, and any increase in penalties if intent to hide can be proved.

There are a variety of potential costs of having violations detected. A fine may be imposed, or a plant shut down. Negative publicity may ensue, as when newspapers publish a "dirty restaurants list."[29] The discovery of below-standard items may slow production.[30] And increased inspection may follow, with all its attendant costs, as the regulatory agency intensifies its scrutiny of problem firms.[31] If all such costs are minimal, of course, the inspectee has little incentive to devote his scarce resources toward hiding deficiencies.

An important determinant of the inspectee's actions is his view of the reasonableness of the regulations. If he believes that the regulations are unjustified, he may hide infractions simply out of resentment, and in other ways hamper the inspection agency. This may be a particular problem when regulations are initially imposed on established production patterns. When FDA authority was extended to cover medical devices, for example, inspectors met unaccustomed resistance from manufacturers. The inspector's prior experience was in the food and drug industries where inspections have long been accepted as part of doing business.[32] A few years ago, when the District of Columbia tried testing automotive exhausts during the regular safety inspections, "people when crazy." Even though it was an experiment, and no one needed to comply, 20% of the motorists refused to let the inspectors near their tailpipes.[33] New regulations, of course, often create substantial short-run problems for producers. The inauguration of the poultry inspection program in the late 1950s led to complaints that it caused interruptions, slowed down dressing lines, and in other ways significantly increased costs.[34]

The Inspectee's Ability to Hide Defects

The ability of the inspectee to hide violations—the cost of hiding violations—largely depends on the nature of the inspected items, the help the inspector receives from third parties, the inspection tactics, and the inspector's legal authority.

What Is Inspected

Some problems are more difficult to hide than others. It is hard to conceal many deficiencies from a motivated automobile safety inspector. Defective windshield wipers, brake lights or turn signals are easily spotted. (Even in a full inspection state, the registered motorist has other options. He may not get his car inspected and try to conceal that fact from the police, or he can purchase a black-market inspection sticker and hope not to be discovered.) Similarly it is quite difficult to hide problems from a competent elevator inspector. This is one situation, moreover, in which the inspector is much more technically knowledgeable than the regulated party. The elevator owner, who generally contracts out for service, probably has little idea of how to conceal a deficiency.

A main reason why it is difficult to hide violations from a concerned elevator or motor vehicle safety inspector is that they are examining pieces of capital equipment. Other inspectors also generally have less trouble monitoring machines. It is easy for the fire marshal to insure that the sprinkler system is properly installed, or for the housing inspector to determine if the boiler is functional or the sink works (or indeed whether the bathroom has a sink), or for the restaurant inspector to observe whether the thermometer in the walk-in refrigerator gives accurate readings. It is more difficult to monitor behavior. It is hard for the fire marshal to know whether exit doors are occasionally obstructed, or for the housing inspector to notice if garbage is sometimes left in the halls, or for the restaurant inspector to guarantee that employees always wash their hands, place their long hair in a net, and keep the can-opener free from dirt.

Conduct requires continuous monitoring if all violations are to be detected. Full inspection of behavior is, of course, prohibitively expensive, so some sort of sample must be taken. In other words, although it is sometimes possible to examine all regulated firms or individuals, they cannot be inspected every minute of every day. So some times must be selected for observation, others omitted. The threat of being caught and penalized supposedly deters much undesirable behavior during the unobserved period—particularly if the inspectee does not know when he will be examined, or misconduct leaves traces. But penalties are usually not severe (nor necessarily should they be), and much illegal behavior is tolerated.

One could argue, of course, that capital equipment also needs continuous examination if deficiencies are to be detected immediately. But there is a crucial difference: a properly working piece of machinery is usually longer lasting than is proper behavior. There is generally not a strong incentive to tamper with the sprinkler system or boiler, or even the refrigerator thermometer. Once the equipment has been installed, the inspectee often has an incentive to maintain it properly. (There are, of course, exceptions. Noteworthy is the possiblity of motorists tampering with emission controls to increase gas mileage.) While deterioration is expected, it occurs somewhat slowly and predictably. By contrast, the inspectee usually has more of an incentive to worsen behavior after the inspector leaves. Or at least there may be less reason to maintain compliant behavior, and it can deteriorate relatively rapidly.

Conduct is difficult to monitor not so much because it can be camouflaged, but because it can often be changed both quickly and inexpensively. If inspection could be undertaken without the knowledge of the regulated firm, it would be much easier to discover violations. But this is rarely the situation. The firm generally knows it is being examined, often has some forewarning, and can alter at least some behavioral patterns. Thus not only is there less than complete or 100% inspection, but the sample taken is systematically biased.

The behavior under immediate observation will most likely be modified. The motorist who spots a police car behind him is unlikely to push his automobile above the speed limit. The restaurant owner will throw away rather than save the food that has fallen to the floor while under the watchful gaze of the inspector. If the government surveyor is present, the nursing home attendant will try to properly bathe, feed and position patients and neither neglect nor mistreat them.

Evidence of prior misbehavior may also be removed if sufficient warning is given of the impending inspection. The motorist slows down when alerted to an approaching speed trap. The firm clears its exit door, the landlord removes the garbage from the hall, and the restaurant cleans its can-opener. When it learns that the OSHA inspector has arrived at the plant gate, the manufacturing firm may even turn off some of its machines.

One way to try to monitor some sorts of behavior is to prescribe recordkeeping for various activities. Thus nursing homes in Massachusetts must maintain a multitude of somewhat detailed accounts, including information on the rehabilitative and social services actually provided as well as nursing plans and medical care evaluations. Inspectors then spend most of their time reviewing these records. Firsthand observation and conversations help verify that there is more than paper compliance—not only that the records are accurate, but that the intent as well as the letter of the law is being satisfied. The record-keeping is time consuming, potentially rigidifying, and serves only as a crude proxy for examining real behavior. But it can force providers into desirable behavioral patterns; it can instill good habits.

Some records that can be examined may already have been made for other purposes. If the incentives for creating these accounts differ from those the regulation gives the inspectee, they can provide the agency a somewhat unbiased source of information. This can make monitoring much easier. An applicant for a mortgage, for example, is generally asked to produce copies of his previous years' tax returns. This record provides the bank with a reliable minimum estimate of past income, for the incentive was to understate earnings to the Internal Revenue Service. Conversely, the IRS could theoretically look at mortgage applications to help it discover unreported income.

For some inspectors, like IRS agents, the difficult task is not really the monitoring of behavior—the IRS is not so concerned with how you make your income, only that you report it—but the tracking down of things. In this instance, the things IRS agents are supposed to find include unreported income and non-reporting people (i.e., moonshiners, non-filers). Certain kinds of income are relatively easy to hide from the IRS, as evidenced by the gigantic estimates of the U.S. "underground economy."

Customs inspectors, immigration officials and policemen are also responsible for tracking things down. Customs officials often cannot detect the carefully hidden illegal item in the enormous volume of legitimate luggage and cargo.[35] Immigration officials have great problems locating illegal aliens residing in the U.S.[36] And the fact that the vast majority of serious crimes are unsolved indicates the difficulty police have in discovering (and helping convict) the guilty parties.[37]

The principal tactic used by the inspectee in such instances is camouflage. Camouflage, however, is difficult if you are too conspicuous. It is harder to smuggle in an elephant than pieces of ivory. It is less easy to get away with murdering one's spouse than a complete stranger. A Teddy Kennedy is more likely to be caught if he cheats on a college exam than is a typical student. A John Kenneth Galbraith cannot escape detection if he fails to report his royalty income.

The inspectee can try to become and remain inconspicuous, or make deficient items indistinguishable from acceptable ones. The marijuana smuggler dresses like a conservative business-man rather than a spaced-out hippie. The bank robber makes no large, unusual purchases. The food processing company's defective eggs are deodorized so they smell like good ones.[38]

Third-party Help

Aside from the inspector and the inspectee, there are other interested people and institutions whom we call "third parties." Third parties can provide information and other aid to the inspector, improving his ability to discover (and certify) infrac-tions. Occasionally, they help the inspectee hide his violations, as when an oncoming car warns the motorist of an approaching speed trap. In some areas, third parties initiate the inspection and play an important part in the entire monitoring process. This is the case with housing inspection, where tenants are usually relied upon to notify the agency of deficiencies. In other areas, such as motor vehicle inspection, third parties play almost no role at all.

It seems useful to distinguish between two categories of third parties: complainers and informers. The complainer is directly and significantly affected by the regulations. He is often en-gaged in an economic transaction with the inspectee, and the regulations were written specifically to protect him. The sincere complainer feels injured by the inspectee's actions that violate the agency's prescriptions. Thus the wholesaler complains if he believes the fresh fruit and vegetables he receives do not meet the stated standard. The diner may complain to the agency about unsanitary restaurant conditions. The employee may complain about unwarranted workplace hazards.

Some housing inspectors investigate only when there are

complaints. They then usually step into the middle of a land-lord-tenant dispute. The tenant gives information to the in-spector, and makes it difficult for the landlord to camouflage violations. But given the small penalties, the landlord may already have little incentive to try to escape detection. A more compelling strategy is often to show good faith by promptly making repairs after infractions have been cited. This usually satisfies the courts and the agency. The tenants, however, may be less content, especially if the improvements are superficial or temporary.

OSHA inspectors are required to seek out and listen to the views of labor. Workers possess large amounts of useful infor-mation, but inspectors are sometimes unsuccessful in their at-tempts to obtain it. Employees may correctly fear management reprisals should they help expose deficiencies. This problem is particularly serious in non-union shops. (Tenants in housing projects are sometimes reluctant to complain for fear of retal-iatory evictions).[39] More information is elicited from workers when they are better protected. Perhaps OSHA's most effective tactic is to provide them with camouflage. The company will be less able to identify offending employees—and thus retaliate against them—if large numbers of workers are individually questioned, all out of earshot of management.

Complaints sometimes initiate restaurant, elevator or weights and measures inspections. The role of complainers is different in housing, OSHA and even nursing home inspections because they are permanently on the scene and can easily provide in-formation during the inspection. That they did not complain officially to the agency does not now preclude face-to-face dis-cussions with the inspector. They not only point out problems, but also lend their own distinct perspective. Such interaction, we shall argue, affects the inspector's incentives, making him less prone to be "captured" by the regulated firm.

Informers are those not directly or significantly affected by the regulations who nonetheless provide useful information to the inspector. Informers include such types as eyewitness by-standers, tipsters, and whistle blowers. FBI agents and police detectives solve crimes not so much from physical clues as from information obtained from knowledgeable parties.[40]

Immigration and customs officials, and even the Internal

Revenue Service receive help from anonymous and paid tipsters. And any regulated firm may have concerned or disgruntled current or former employees who become whistle blowers, exposing company infractions.

Regulatory agencies that inspect products usually do not pay informers. They may, however, receive useful information from whistle blowers, and they often try to protect their anonymity. The Coast Guard, for example, which regularly inspects for vessel safety, makes it a court martial offense for any of its officers to divulge the name of an informant.[41]

A few agencies do reward tipsters. The Bureau of Customs has an established bounty system under which it pays up to 25% of the amount collected, to a maximum of $50,000.[42] The IRS pays informers—though not a great deal. In 1979, fewer than 500 citizens received an average of $650 each for usable information.[43] In general, it seems that financial rewards for third-party information are offered primarily in areas where violations are viewed as potentially criminal rather than civil. Additionally, it is often easier to have a system of paying informants when the agency itself is in the business of collecting money.

It is illuminating to contrast the role third parties play in affecting the ability of management to hide infractions in nuclear plant versus nursing home inspections. Nuclear inspectors can receive some information from third parties, particularly workers. A good resident inspector, similar to a foot patrolman on an established beat, can gain people's respect and confidence, and be the recipient of useful information.

But a nuclear inspector talks almost exclusively to industry personnel. By comparison, the nursing home surveyor not only monitors workers, but also examines and visits with patients. Moreover, many outsiders—relatives, friends, doctors, social workers, private nurses—spend time in the nursing home. And many violations are readily apparent, even to an untrained eye. Visitors can and do see crowded conditions, dirt, inadequate attention, etc. In some sense the home is under constant observation. The setting is vastly different for a nuclear facility, which rarely has outsiders wandering through, and where only experts can recognize most deficiencies.

A nursing home surveyor can thus rely on third-party help to a far greater degree than can a nuclear inspector. In addition, of course, the presence of third parties means that the inspector's behavior is also under greater scrutiny. This relieves the regulatory agency of some of the burden involved in monitoring its own inspection force.

Tactics

The relationship between inspector and inspectee is not one of pure conflict. They are not playing a zero-sum game. In many facets of their interaction, cooperation can make both better off. The inspector can be obnoxious, picayune and obdurate and impose great costs on the regulated firm, but he gains little from acting in this manner. Indeed, he will undoubtedly be made worse off, for the regulated firm can retaliate in a variety of ways, imposing costs on him. Most of the time both parties benefit by acting civilly and reasonably, displaying some respect and regard for the other. (Even warring nations, whose relationship is far more hostile and antagonistic than that between inspector and inspectee, benefit from some cooperation. All are generally better off if they respect white flags, feed prisoners and refrain from germ warfare. Sometimes, of course, it is advantageous for one side to break such implicit and explicit agreements.)

While there are some elements of cooperation, the inspector and inspectee often have opposing goals. Each can employ a variety of tactics in his attempt to discover violations, or to hide them. This section briefly discusses a few of these tactics.

The agency must decide whether or not to announce inspections. Sometimes this is not an issue. The inspectee effectively selects the time for border inspections by customs and immigration officials. The timing of export grain and fruit inspections is largely detemined by when the produce is ready for shipment.

Surprise inspections are called for if violations can rapidly, and momentarily, be corrected. This is often the case when behavior is being monitored. Speed traps are concealed. Restaurant inspections are generally unannounced and occur frequently.

There are some situations when surprise is desirable, but the inspection takes a substantial amount of time. Nursing home, nuclear and OSHA health inspections may be included in this category. The regulated firm can be surprised only at the beginning of the inspection. The second day's visit is fully expected. Moreover, the first day's opening conference may last an hour or so, possibly long enough for the company to take steps to eliminate some infractions.

It may be difficult to truly surprise the inspectee. In air quality monitoring, the desire of the EPA to coordinate and ease relations with state officials increases the likelihood that the source will learn the inspection date.[44] Even if the exact schedule is a well kept secret, the regulated firm may anticipate the inspector's arrival. The fact that nursing home certification is usually required annually means the facility can deduce the approximate time of the next site visit.

Surprise inspections sometimes decrease the ability of the inspector to do his job. This can be the situation if the inspectee needs to provide certain materials, information or help. In a nursing home survey we went on, the nutritionist's day off made it difficult to adequately evaluate the appropriateness of the patients' diet. Housing inspectors usually need someone to be home so they can get into the apartment. Elevator inspectors rely on the aid of service personnel. Their examinations are thus almost always coordinated in advance.

Some inspections are normally scheduled. Certain aspects of Coast Guard vessel inspections should be performed only while the ship is drydocked. Similarly, routine Federal Aviation Administration (FAA) inspections of airplanes take a certain amount of time, and could overly disrupt airline schedules were they not coordinated in advance. Surprise inspections are less critical in these areas since it is difficult to camouflage violations, and improvements are usually long lasting.

Whether to have surprise inspections is an aspect of the broader question of what sorts of information the agency should conceal from, make available to, or publicize for the inspectee. Should the inspectee be told, or be able to determine, how often inspections will occur, what items the inspector will examine, what he will focus upon, his ability to discover infractions, the punishment that will actually be imposed, etc.?

Some agencies are more open than others about such issues. The Coast Guard, National Aeronautics and Space Administration (NASA) and the FAA furnish copies of inspection check-lists.[45] Many agencies also announce their priorities, explaining which items they believe are more important. This tactic is often used when complete compliance with the regulations is not expected. The firm then is more apt to satisfy those specific requirements emphasized by the agency, and perhaps is less likely to satisfy others. If the inspector also focuses on the emphasized items, he may discover only a small percentage of the total violations. But the agency may be better able to fulfill its actual mission.

In Massachusetts, nursing home surveyors rate each facility on 627 items. Providers are not expected to receive perfect scores, and almost none is found to be flawless. The survey ratings form the basis not only for the Pass/Fail certification decision, but also for the determination of reimbursement levels. The regulatory agency does not consider all 627 items equally important, and it gives them different emphasis. Not only is the surveyor checklist available to providers, but so are the agency weightings for each item.[46]

Some agencies seem more concerned with hiding certain of their detection efforts. The IRS, for example, tries to conceal the exact method it uses for selecting which returns to audit.[47] Customs officials do not divulge their criteria for choosing which individuals to search intensively. Immigration does not tell exactly where along the border it deploys its resources. Speed traps are often moved so motorists cannot tell where they will be observed.

While the entire population of items may be superficially examined, these agencies carefully scrutinize only a small sample. Their selection processes are different, and, they believe, better than simple randomization. They want to keep their method secret. Should their decision rules become public, they think the inspectees would use such information, not to move substantially into conformance with the requirements, but to make noncompliance more difficult to detect.

Other reasons may underlie the secrecy of these agencies. In a customs, immigration or police examination for example, the inspector-inspectee relationship is usually a fleeting one. There may not be a future opportunity to punish current misbehavior.

What information to provide to inspectees is generally an

agency-wide decision made at the management level. The individual inspector is also able to select among various inspection techniques. His mandate is narrower and his strategems are often specifically designed to help him discover infractions. The inspectee, of course, can use a variety of ploys to make that particular task more difficult.

One approach for an Internal Revenue Service agent is to act tough, impersonal, crisp and polished. The agent then tries to capitalize on the initial nervousness of the taxpayer and increase his disorientation by removing him from familiar surroundings. An alternative is to try to put the taxpayer at ease. Meet in his home—where the agent can better evaluate his lifestyle. A friendly demeanor, a relaxed conversation and the taxpayer may lower his defenses and inadvertently blurt out incriminating information.

The taxpayer has some maneuvers of his own. Time is generally on his side, for delay lengthens the time he has use of the government's money. More important, agents believe their performance is evaluated on the basis of cases closed and the amount of money collected.[48] They are thus generally pressed for time. They do not like to have cases continued. The agent may ask the taxpayer to bring certain documentation to the audit. If, during the proceedings, something else piques his interest, he can ask about those items. A response of "but I didn't bring the documentation for that—you didn't ask for it" is often an effective tactic. The agent definitely does not want it to appear that his own possible negligence is the principle cause for continuance.[49]

Grain firms have also taken advantage of the inspector's time constraints. The actual examination of U.S. grain is performed by licensed, often private, inspectors, subject to oversight by federal supervisors. Department of Agriculture personnel believe that, during peak inspection periods, the regulated firms have sometimes requested frivolous appeal inspections—which require the presence of USDA officials—to limit their availability for the more important supervisory functions.[50]

The regulated companies have also acted to make the primary inspector's job more difficult. When grain is exported, the vessel storage area must be examined to ensure that it is not wet, dirty, contaminated or infested. Stowage inspections can usually be made quickly, but they must be performed before the cargo is loaded. A GAO investigation uncovered an instance where the hold had been fumigated shortly before the arrival of the inspector. Since it was still noxious, he gave it a superficial inspection, never venturing inside for the specified examination.[51]

Because inspectors were often accommodating, the companies could usually guarantee a cursory examination by making a good inspection costly, not only for the inspector but also for themselves. Thus one firm loaded its cargo before receiving approval. The enormous expense of unloading induced the inspector to forego the normal requirements.[52]

Legal Authority

The inspector's legal authority obviously affects his ability to discover violations and the inspectee's ability to hide them. There is usually a trade-off between public protection and individual liberty, between legitimate government concern and personal or property rights.

The ability to search and seize can be of great benefit to the inspector. A customs inspector has considerable power in this regard. He has the right to examine the personal property of a traveler without a warrant or even probable cause.[53] Strip searches are permitted if there is "reasonable suspicion," such as aroused by a tip or suspicious behavior.[54] By contrast, an OSHA inspector can be kept out of the plant unless he obtains a search warrant. This increases his administrative burden as well as decreasing his ability to undertake surprise inspections.[55]

Immigration inspectors currently have little chance of finding the majority of illegal aliens residing in this country. Of major help to INS would be the requirement that every person carry a national identity card and show it to get a job or to an immigration official or policeman upon request. But the ID-card idea has been rejected in the past as an infringement on citizens' freedom and civil rights.[56]

The Internal Revenue Service can gain access to the taxpayer's bank records. Without this power, it would be much more difficult to either detect or prove the existence of much unreported income.[57] By contrast, FDA officials complain that their inability to demand technical and production records from food facilities seriously decreases their effectiveness.[58]

Police claim their ability to catch speeders is improved if radar detectors are outlawed. Whether or not they can extract confessions is influenced by the degree to which they are allowed to browbeat suspects. The legal status of wiretapping affects their ability to bring to justice underworld leaders who order, but do not carry out, criminal activities.

Arms control treaty negotiations often flounder over the issue of inspection. The ability of each country to monitor the others depends in large part on what the inspectors are permitted to do. Can they go into manufacturing plants? Are they allowed to roam through military installations? Can they give lie-detector tests to knowledgeable personnel? Increasing the legal authority of arms inspectors may result in a small decrease in traditional liberties; more importantly, it limits each nation's ability to protect and safeguard what it may consider to be legitimate military secrets.[59]

Power of the Inspector to Gain Compliance with the Regulations

The focus of this book is on whether or not the inspector will be a good reporter. Reporting, of course, may be just one aspect of the inspector's job. The basic function of inspection is to help the agency accomplish its mission, to gain compliance with the important regulations at low social cost. This section does not look at the inspector's ability to discover violations. Instead, it briefly discusses some of the other factors which affect the inspector's ability to gain inspectee compliance. The adversarial nature of the inspector-inspectee relationship is apparent throughout.

The Need to Document Noncompliance

The inspector not only needs to detect violations, he must also certify them. For some inspectors, this creates little additional burden. Motor vehicle, apple, and fruit and vegetable inspectors need only follow customary procedures and then state their findings and observations. Should the inspectee disagree with the analysis, he may request and receive a re-inspection. A product, rather than behavior, is being examined in these instances. It is considered acceptable for the item to be improved in the interval before reinspection, for the important issue is its final quality or safety. Moreover, there is not normally a fine or criminal penalty for possessing the defective product— only for possibly using, selling or passing it off as a higher grade.

Housing inspectors also have few documentation requirements. But their task is sometimes difficult because they must confirm alleged violations by observing them first hand. All the tenants may complain of rats in the building. The inspector, however, must see the rats himself, or at least find evidence of current infestation. Or a tenant may claim that the roof leaks. While water stains are apparent, it may not be obvious how old the stains are. When contacted, the landlord explains that he has recently patched the roof. To cite an infraction, the inspector must have first-hand information that the repair was inadequate—such as actually seeing the roof leak.[60]

In citing traffic enforcement, policemen are like most inspectors in that they must rely only on their own evidence. When the 55 mph speed limit first took effect, police were deluged with calls from citizens giving the license plate numbers of motorists observed speeding. Such information may not be used by the police. But once an officer himself spots a violation, little documentation is necessary. The presumption is that the police act correctly and competently in the performance of their normal duties. If there is any question about the citation, the burden of proof is on the motorist to prove his innocence.

Somewhat similarly, the burden of proof is on the taxpayer to document any claimed deductions. By contrast, it is up to the IRS to demonstrate the existence of unreported income. Thus it is generally easier for an audited taxpayer to hide income than to successfully claim unwarranted deductions.

Nursing home surveyors in Massachusetts must give a written justification for every citation. But they need not provide proof that a violation in fact occurred. Their expert observation is generally considered sufficient. A different quantum of evidence is required from the OSHA inspector. He is often expected to create a permanent objective record of cited infractions.

Therefore, he normally carries a camera with him. Challenges to OSHA decisions forced the agency to adopt a legalistic approach and to place great emphasis on the documentation of findings. This in turn put an additional burden on the OSHA inspector. It gave him less power, and less leverage with which to coerce workplace improvements.

In all Pass/Fail inspections there can be both Type I and Type II errors—false positives and false negatives. A Type II error occurs when defective items are classified as acceptable, or when violations are not cited. The easier it is to conceal infractions, and the more stringent the documentation requirements, the more likely it is that there will be Type II errors. A Type I error is the rejection of acceptable terms, or the citation of infractions when there are in fact no defects. Strict documentation requirements decrease this problem.

The Enforcement Power of the Inspector

Increased agency enforcement will generally cause the regulatee to try to have fewer and less serious violations reported. They may accomplish this by spending more resources (or using their resources more efficiently) to hide violations, to capture inspectors, or to improve conditions. The last option is, naturally, the one preferred by the agency.

Some inspectors are given enforcement powers. A principal benefit this brings the agency is to speed up the enforcement process. It can also enhance the inspector's self-image and thereby his overall performance. There are problems, of course, from management's perspective, in that it can give low level personnel too much authority, making it difficult for the agency to pursue a consistent, coherent and even-handed policy.

Many inspectors have no enforcement power. These include housing inspectors, motor vehicle inspectors, OSHA inspectors and nursing home surveyors. All enforcement at the FDA comes from central headquarters. In the EPA, while enforcement is primarily carried out by field personnel, the actual inspector has no enforcement authority.[61]

The inspectors who possess enforcement power usually cannot impose civil or criminal sanctions. Instead their authority permits them to condemn products and halt operations. Apple, and fruit and vegetable inspectors not only grade products, but they can also condemn them. Weights and measures inspectors have the authority to stop equipment from being used. Elevator inspectors also have shut down power. Restaurant inspectors can close facilities.

Building and Coast Guard inspectors possess authority since their approval is required before business operations can begin. The builder needs a "certificate of occupancy" before anyone can legally inhabit the premises. The shipowner must have a "certificate of inspection" before the vessel sails. So costly is a maritime delay that an inspector's decision has never been formally challenged. It has always been much less expensive to make the required improvements than to go through a time consuming appeals procedure.[62]

The Negotiation Position

Inspectors have discretionary power. For example, there is typically a grey area in any regulation and choices must be made about whether or not to cite marginal infractions. An interesting interpretation of the building code was made recently by a New Jersey inspector who decided to find a tree house in violation of the rule limiting the number of dwelling units to one per lot.[63]

The inspector clearly has discretionary power when he haggles with the inspectee. And it often makes sense for him to do so. Certification of violations and the legal enforcement of regulations can be costly to the agency. Documentation may be difficult. Court appearances are time consuming. Additionally, formal enforcement does not always result in either an effective or swift improvement in conditions. These are some of the reasons an inspector may find it prudent to haggle. He may say, "If you don't challenge my judgment concerning this violation, I won't cite that one." Or, "If you correct this problem right away, I won't report it." While IRS agents and OSHA inspectors are instructed not to bargain, it is common knowledge that both do. And often for good reason. It is certainly possible for an ideal inspector not to be a completely accurate reporter.[64]

The inspector's incentive to strike a "good bargain" if he

negotiates is discussed in the next chapter. His ability to do so depends on a variety of factors. One is the documentation requirements imposed on him by the agency and the courts. Where documentation is difficult, a good inspector may want to bargain, but may be in a relatively poor negotiating position. Another factor is the enforcement power of the agency. If it is weak, the inspector again has little leverage.

The inspector may have recourse to a variety of threats that can improve the final outcome of negotiations. The IRS agent, for example, threatens to scrutinize other deductions or even previous years' tax returns if his decisions are not accepted. Or he mentions that he could be forced to question the taxpayer's business associates—informing them that the individual is under IRS investigation, possibly damaging his reputation.[65] Such tactics may not improve the agent's ability to detect violations. But they make it less likely that the inspected party will contest his decisions.

IV. SUMMARY

Regulations are generally created to impose costs on socially undesirable behavior. This makes inspection adversarial. The inspector's principal task is to search for violations. The regulatee usually prefers that none of his own be discovered. There is thus conflict, and room for strategic behavior.

There are two principal ways that the individual inspectee can try to decrease the chance his infractions will be detected and reported. The first is to "capture" the inspector, decreasing his incentive to seek out or cite violations. The second method is to hide the problems. One means of concealment is to alter the conditions prevailing during the inspection period. This can be particularly effective if behavior is being monitored, and there is advance warning of the impending inspection. An alternative means of hiding violations is through camouflage. The inspectee does not eliminate the infractions, even momentarily. He simply conceals them from the inspector's purview.

Many factors affect the ability of inspectors to discover most serious violations. A few of these, such as the in-house training of the inspectors, are under the control of the agency. The agency management, however, has only limited influence on many other factors, such as the number of inspectees, the nature of the inspected item and the availability of third-party help. The thesis of this book is that such features of the inspection environment largely determine the individual inspector's and the inspection force's ability to do a good job.

For agencies such as OSHA and INS, however they are managed, it is virtually impossible for the inspection force to do a good job. The OSHA staff is too small to examine more than a small percentage of all workplaces. Moreover, given the great diversity of manufacturing processes and locations, it is difficult for the individual inspector to be an expert at every plant he enters. Immigration inspectors lack the resources for effective surveillance of the entire U.S. border. Additionally they have neither enough information nor authority to pose more than a minor threat to the illegal aliens currently residing here.

In contrast, motor vehicle inspectors clearly have the ability to discover most important infractions. In many locations they regularly examine close to 100% of the regulated universe, and violations are both difficult to disguise and easy to detect. The question, however, is not whether they *can* do a good job but whether they *will* perform competently. The inspector's incentives are examined in the next chapter.

NOTES AND REFERENCES

1. S. R. Thompson et al. "Evaluation of Inspection and Enforcement Programs of Other Regulatory, Safety and Professional Organizations," Teknekron, Inc., Washington, DC, 1978. Prepared for the Office of Inspection and Enforcement, U.S. Nuclear Regulatory Commission, Contract #NRC-05-77-064, Vol.1, p.55.

2. Interviews with Massachusetts Nursing Home Survey Teams, 1981.

3. Paul Strassels and Robert Wool, *All you Need to Know About the IRS* (New York, NY: Random House, 1979) p. 95.

4. Merle Broberg, "A Study of a Performance Control System as an Indicator of Organizational Goals: The Housekeeping Inspection System of the Philadelphia Housing Authority." Ph.D. dissertation in Sociology, American University, 1969.

5. College of American Pathologists Accreditation and Nuclear Energy Liability Property Insurance Sales, S. R. Thompson et al. "Evaluation of Inspection and Enforcement Programs of Other Regulatory Safety and

Professional Organizations," Teknekron, Inc., Washington, DC, 1978, Vol.2, pp.75,153.

6. California Radiologic Health Section of the State Health Department and Illinois Department of Public Health, *ibid.*, pp.183,102.

7. James Cassedy, "Applied Microscopy and American Pork Diplomacy: Charles Wardell Stiles in Germany: 1898-1899," *ISIS*, Vol. 62, Spring 1971, pp.15-16.

8. Thompson et al., *op. cit.*, Vol. 2, pp.87,36.

9. *Ibid.*, p.129.

10. Interview with restaurant inspector.

11. Thompson et al., *op. cit.*, Vol.1, p.80.

12. *Ibid.*, p.85.

13. "Now OSHA Must Justify Its Inspection Targets," *Business Week*, April 9, 1979, p.64.

Thompson et al., *op. cit.*, Vol.2, p.156.

14. *Ibid.*, pp.77,59.

15. O. Sussman, "New Law's Dangerous Deceit," *Nation's Business*, May 1968, pp.34-37.

16. Interviews with OSHA inspectors.

17. General Accounting Office, "Placing Resident Inspectors at Nuclear Power Plant Sites: Is it Working?" November 15, 1979, p.1.

18. *Ibid.*, p. 12.

19. General Accounting Office, "Improvements Still Needed in Coal Mine Dust-Sampling Program and Penalty Assessments and Collections," December 1975, p.24.

20. Jeffrey Lenorovitz, "DC-9 Tailcone Loss Spurs Aft Bulkhead Inspections," *Aviation Week and Space Technology*, September 24, 1979.

21. Scott Henderson, "Thefts, Losses Pose Perennial Headache," *Harvard Independent*, March 6-12, 1980, p.5.

22. J. Tedesco, "Let the Camera Be Your Eyes," *The American City*, January 1979, p.85.

23. Leigh Montville, "A Peek Inside of the Room TV Never Will See," *The Boston Globe*, February 13, 1980, p.25.

24. Jane Bryant Quinn, "The Income Tax Cheat May Be Finding It Easier," *The Boston Globe*, April 6, 1981, p.19.

Strassels and Wool, *op. cit.*, pp.46-47.

25. Neil Ulman, "Navigation and the Troubled Tankers," *The Wall Street Journal*, February 15, 1977, p.22.

26. Gerald Leighton and London Douglas, *The Meat Industry and Meat Inspection* (London, England: Educational Book Co.) 1910.

27. General Accounting Office, "Prospects Dim for Effectively Enforcing Immigration Laws," November 5, 1980, p.1.

28. *Ibid.*, pp.1-26.

29. "18 Food Places Called Health Code Violators," *The New York Times*, July 19, 1981, p.27.

30. B. Hardy, "Poultry Men Feel Inspection Pinch," *Farm Journal*, Vol.83, May 1959, pp.72-73.

31. "Manufacturer to Monitor DC-10 Engine/Wing Pylon," *Aviation Week*, July 23, 1979, p.27. Airlines hoped a series of uneventful inspections would lead to a stretching by the FAA of the length of time (intervals) between inspections.
32. Interview with former FDA inspector.
33. Phil McCombs, "Emissions Testing Coming: 350,000 Cars Will Flunk," *The Washington Post*, July 8, 1981, p.1.
34. Hardy, *op. cit.*
35. General Accounting Office, "Heroin Being Smuggled into New York City Successfully," Bureau of Customs, Department of the Treasury, Bureau of Narcotics - Dangerous Drugs, B-164031(2), December 7, 1972. "It is unrealistic to expect Customs inspections to prevent most heroin from being smuggled into the United States," p.1.
36. "General Accounting Office, "Prospects Dim for Effectively Enforcing Immigration Laws," November 5, 1980, p.19.
37. "Only 25% of the burglaries known to the police in 1965 were solved, and many burglaries were not reported to the police."
"Only 20% of reported major larcenies are solved, and the solution rate for minor ones is considerably lower."
President's Commission of Law Enforcement and Administration of Justice, *The Challenge of Crime in a Free Society* (Avon Books, 1968) pp.63-64.
38. Egg Product Inspection Act, 91st Congress, 2nd Session, December 3, 1970. Report #91-1670.
39. Jackson Diehl, "Tired of Complaining: Tenants Lose Hope," *The Washington Post*, July 23, 1979, p.A19, Col.1.
"And above all, they (tenants) fear that if they complain, they will be giving the management a reason to evict them."
40. Arthur Niederhoffer, et al. *The Ambivalent Force: Perspectives on the Police* (San Francisco, Rinehart Press, 1973).
41. Thompson et al. *op. cit.*, Vol.1, pp.107-108.
None of the twenty regulatory agencies surveyed for this volume provided compensation for informers.
42. Stephen Birnbaum, "Customary Procedures," *Playboy*, June 1980, p.59.
43. Strassels and Wool, *op. cit.*, p.22.
44. Thompson et al. *op. cit.*, Vol.2, p.173.
45. *Ibid.*, pp.111,141,194.
46. Interviews of Nursing Home Surveyors.
47. Strassels and Wool, *op. cit.*, pp.36-37.
48. General Accounting Office, "How the Internal Revenue Service Selects Individual Income Tax Returns for Audit," November 1976, p.74.
49. Strassels and Wool, *op. cit.*, Chapter 11.
50. General Accounting Office, "Supplemental Information on Assessment of the National Grain Inspection System," Department of Agriculture, CED-76-132, July 1976, Attachment I, p.33.
51. *Ibid.*, Attachment II, p.35.
52. *Ibid.*
53. Birnbaum, *op. cit.*

54. Rosemary Alexander, "Unwarranted Power at the Border: the Intrusive Body Search," *Southwestern Law Journal*, Vol.32, November 1978, pp.1005-26.

55. "Bill Vindicated - Inspectors May Need Warrants," *Time* 111:16, June 5, 1978, p.16.

56. General Accounting Office, "Prospects Dim for Effectively Enforcing Immigration Laws," November 5, 1980, p.10.

57. Strassels and Wool, *op. cit.*, p.14.

58. Thompson et al. *op. cit.*, Vol.2, p.46.

59. Louis Henkin, "Arms Inspection and the Constitution," *Bulletin of the Atomic Scientists*, Vol.XV, No.5, May, 1957, p.197.

60. Pietro Nivola, "A Municipal Agency: A Study of Housing Inspection in Boston," Ph.D. Dissertation, Harvard University, May 1976, Chapter IV.

61. Thompson et al. *op. cit.*, Vol.2, pp.41-42,55,174.

62. *Ibid.*, Vol.1, p.63.

63. Since the tree house had been built by a contractor and was viewed as an annoyance by neighbors, the inspector's ruling was upheld in the courts. The owners of the tree house have subsequently obtained a satisfactory permit and inspection. (Interview with a New Jersey building official, 1982.)

64. Strassels and Wool, *op. cit.*, p.142.

Interviews with OSHA inspectors.

65. Henry Rothblatt, "Income Tax Evasion: Dealing with the IRS Special Agents and Prosecution," *Criminal Law Bulletin*, June 1974, 10:pp.437-442.

Chapter III

The Inspector's Incentives

During the winter in New York City, the 300 housing inspectors assigned to the heat squad are expected to respond to 3,500 daily complaints from apartment dwellers. Inspectors tell grisly anecdotes of mad dogs, piles of garbage, squealing rats, and the relentless cold.

New York Times, February 17, 1979

An in-house nuclear inspector told CBS News that workers threatened his life and beat up his colleague when they pointed out construction flaws. He claimed that for months no one dared monitor the construction, and that the inspectors simply falsified forms.

Newsweek, October 1979

Chapter II focused on the ability of inspectors to effectively monitor the regulatee. The question was, "Can they do a good job?" This chapter examines the inspectors' incentives. The issue is, "Will they try to do a good job?" Since there are many facets to their work, they could, of course, try hard in some areas but not in others. For example, an inspector might be very interested in some infractions and direct his attention toward discovering these, while at the same time overlooking other violations. While our emphasis is on the inspector's overall inclination to perform as capably as he can, we also discuss his incentive to do better in some areas than in others.

Many factors affect the inspector's incentive to do a good job.

One broad influence present in virtually every kind of employment may be termed "Work Satisfaction."[1] We separate Work Satisfaction into two components: the quality of the job and its perceived usefulness. Issues of Job Quality include whether the inspector's work is interesting or dull, the degree of autonomy he has, how much he uses his training, and whether a capable performance places him in additional physical danger. Job Usefulness involves a belief that the mission of the employer is worthwhile, and that the worker contributes toward its fulfillment.

A second general influence on the worker's incentive to do a good job is Performance Evaluation. Can his performance be monitored? In other words, is good (or bad) work noticed? And if it is, what kinds of rewards (or punishments) are meted out? Where worker performance can be evaluated and good work is rewarded, the employee is more likely to try his best. If only some aspects of the work can be monitored, or only some are recorded, the worker will naturally tend to focus on these areas.

What distinguishes an inspector's work from most other employment is its adversarial nature. The typical inspector meets face-to-face with the regulatee, who probably prefers that the inspection not take place, and almost definitely desires that few violations be discovered. Workers in other fields do not normally encounter such conflict. Not everyone is happy if the inspector does a good job—often this displeases the inspectee. And being a tough examiner, citing every possible violation, is sure to upset him.

The distinct possibility exists that the inspector will be "captured," becoming fearful of the regulatee or sympathetic with his plight. In other words, if the inspector does not already identify with the inspectee he may be won over by a variety of arguments, favors, threats and bribes. There is certainly pressure on the inspector to make Type II rather than Type I errors, to be lax in searching out and reporting violations.

The converse of capture is extortion. Unlike most employees, the inspector may have the opportunity to demand tribute. The regulatee may be forced to make illegal payments to the inspector for performing competent—timely and not excessively strict—examinations. We call the possibility of either capture or extortion the inspector's "moral hazard" problem.

I. WORK SATISFACTION

Job Quality

The inspector's work can run the gamut from exceedingly dull to dangerous and exciting. Like most jobs, however, it typically becomes routine. Although there have been famous literary inspectors—Nathaniel Hawthorne and Herman Melville, for example, both worked for the customs service[2]—there are few reliable descriptions that effectively give a feel for the inspector's daily work.[3] There is, of course, agency propaganda. The day-to-day travail of the FDA inspector is glowingly depicted as:

> an exciting drama in which he and others apply scientific skills and common sense to protection of the public health. The inspector's work is seldom routine as he examines the sanitary conditions in food, drug or cosmetic establishments...reviews analytical work performed by scientists...makes on-the-spot examinations...interviews consumers, industry executives, production and research chemists, merchants and others.[4]

In our interviews with FDA inspectors, none mentioned the drama. The phrase used to describe the work was, "It's a job."

At the other extreme, the popular press gives grim descriptions of the inspector's work—such as the housing inspector's grisly anecdotes of piles of garbage, mad dogs and squealing rats.[5] INS patrolmen "face nightly volleys of rocks, bottles and sometimes gunfire from across the chain-link fence that separates Mexico and the United States."[6] The federal meat inspector:

> works under extremely unpleasant if not nauseating conditions. Most meat-process plants are old, hot, noisy and noisome. The constant sight and smell of rent flesh, blood, entrails, and offal are sensuous assaults to which the inspector may grow accustomed, but never immune. Twelve-hour work days are common.

The inspector must often cover many 'houses' in a circuit, traveling from plant to plant at some distance and at odd hours.[7]

Whether the inspector's work is exciting or grisly or dull interests us here only as it affects his incentives to do a good job. The presumption is that, the more interesting the work, the higher the quality of working conditions, the more likely it is that the inspector—or any employee—will perform up to his capabilities.

Many other factors affect the satisfaction the inspector feels in his position. One is the degree to which he can use his training. Auto mechanics and policemen typically employ few of their skills while acting as motor vehicle inspectors and traffic patrolmen. Additionally, much of such inspection work is tedious and dull. By contrast, elevator inspectors, IRS agents and nursing home surveyors put to use a large portion of their training and expertise.

The inspector's autonomy is another factor influencing his perception of job quality. John Kenneth Galbraith in the *New Industrial State*[8] argues that the two principal concerns of the "technostructure" are salary and autonomy. Most workers set similar store in these twin goals. The inspector's autonomy, however, is limited by the regulatory agency's need to exert direction and control to ensure competence, fairness and consistency in its enforcement efforts. Detailed procedures are thus spelled out and specific routines established. The problem is that while this may raise the performance of the least capable inspectors and make examinations more uniform, the decrease in inspector autonomy can lower morale and inhibit innovation.[9] Increased management control can, in other words, decrease the good inspector's incentive to do an excellent job. OSHA health inspectors have complained that the agency has inadvisedly attempted to transform the "art" of inspection into a mundane, cookbook exercise, diminishing the inspector's job satisfaction (and also lessening his ability to perform a useful examination). Instead of increasing bureaucratic control over worker discretion, some argue that it would be preferable to hire more qualified inspectors and give them better training.[10]

IRS agents are one group of inspectors reputed to be generally satisfied with their work.[11] They have little cause not to do their best. The job is white collar, with little physical danger. The pay is reasonable. The work is often interesting; IRS agents are described as "happy men when talking about taxes."[12] They are, moreover, one of the few types of inspectors who are usually the center of attention at cocktail parties.

IRS agents have some expertise, and they get to use it. They possess significant discretion and authority. They make numerous decisions, often involving substantial sums of money. Additionally, they believe in the mission of the IRS and in their contribution to it.[13] And, as we shall argue, that can really matter.

Job Usefulness

Agency Mission

> The inspector is a busy and dedicated individual who has the personal satisfaction of doing a job recognized by everyone as being important and necessary to the health and welfare of the Nation.[14]

This glorified description of the FDA inspector correctly implies that a critical influence on an inspector's incentives is whether he believes his work is useful. That generally requires that the agency itself be perceived as performing a worthwhile function.

There is consensus concerning the benefits of effective enforcement by many regulatory agencies. Few questions are raised about the role the elevator, restaurant or weights and measures inspector should play in protecting the well-being of society.

As a nation, we display much more ambivalence about our housing and immigration laws. Strict inspection and enforcement of the housing codes could be disastrous. Forced upgrading of blocks of substandard apartments can lead to the raising of rents and the pushing out of the poor. Piecemeal attempts to ameliorate conditions can hasten abandonments since the cost of required improvements will generally exceed the increase in property values. Thus housing authorities, and the courts, are usually not zealous enforcers of the codes. The agency will try to ensure quick correction of certain infractions, such as inadequate heat, but in general, the housing inspector cannot gain much satisfaction from viewing the ultimate results of his work.[15]

The situation encountered by INS inspectors may be worse. The national attitude toward illegal immigration is, at best, equivocal.[16] The cause of most unlawful entry into the United States is the desire for better economic conditions. Yet while many voice their concern over the increasing numbers of illegal aliens, there are still no federal sanctions against employers who willingly hire them. While eleven states have enacted laws prohibiting the employment of illegal aliens, the GAO could discover only one case in which any punishment was meted out; the defendant was fined 250 dollars.[17]

Most illegal aliens are simply sent back to their native country without penalty. Thus border patrolmen may arrest some of the same people night after night. The job can become quite frustrating. Selective enforcement of the regulations can also undermine morale. Inspectors are sometimes told that for the sake of "community relations" they must avoid certain ranches, hotels and restaurants that are known to employ illegal aliens. Unofficial exemptions may also be given to women working unlawfully as domestics.[18]

Double standards in enforcement and the lack of public support for their mission decreases the inspector's incentive to perform up to his capability. One senior INS official told a *New York Times* reporter: "Everybody is basically dissatisfied. There's bound to be corruption." Said another: "It gets to the point where, after a while, there's just no enforcement. An awful lot of officers just give up."[19]

Specific Task

An inspector will perform better when he thinks not only that the agency has laudatory goals, but also that he contributes to their achievement. He must both believe in the value of his own task, and have the ability to do it well.

An inspector can become frustrated if he is expected to enforce silly or unfair regulations. In the 1970s, for example, some of the regulations covering large container ships were outdated, holdovers from the clipper ship era. While the Coast Guard insisted on regular inspections of the magnetic compass and ship's whistle, the regulations did not mention the radar, the most vital equipment for avoiding collisions.[20] OSHA standards have been censured since, following the Congressional mandate, they were adopted en masse in 1971. The National Federation of Independent Business, Inc., for example, claimed OSHA was often regarded as a "meddlesome, nit-picking agency" owing to its enforcement of "obsolete, insignificant and irrelevant" standards.[21]

The recent strip mine regulations, created to protect the environment, were imposed so quickly that there was insufficient time to explain the law. Inspectors were instructed to hand out violations, although the regulations had not been adequately interpreted for the mine owners. This was a particular problem given the vague language in many of the standards. Different construction of the same regulations by the federal and state governments added to the confusion. This was not the type of situation designed to win friends for the regulatory agencies.[22] It did effectively decrease inspector morale.

No regulations are perfect. A study involving three regulatory agencies found that "almost everyone whose plant or worksite was inspected (during the time the case studies were conducted) identified individual regulations they felt were foolish or cumbersome,"[23] and the inspectors could each "cite instances when the people being regulated were justified in expressing anger or resentment about the regulatory process in general or the inspection process in particular."[24]

The inspector can also become frustrated if his specific task is worthwhile but he is unable to accomplish it. The employer can require an OSHA compliance officer to obtain a search warrant before he enters the plant. Surprise inspections can thus effectively be prevented. Since backlogs may exist at the solicitor's office, the inspector can be stalled for some time. Even without the need for a warrant, the compliance officer can detect only a small percentage of all violations. Many infractions are transitory, and the OSHA inspector can visit the plant but rarely. Indeed, he is usually told that he arrived at an atypical time, that if he had come a day earlier, he would have seen significantly different conditions. The OSHA inspector also lacks enforcement authority. He does not determine penalties, and he cannot close down even one phase of an operation although it may present a clear and immediate danger to life and limb.[25]

Attitude improves, and thus performance, if the inspector becomes convinced that his contribution matters, if he can improve the situation and is able to do so. One way he may contribute is by directly helping the inspectee, providing positive suggestions rather than merely negative sanctions. Supplying useful advice can also help transform an adversarial confrontation into a more relaxed, cooperative proceeding. This is an argument for:

> permitting substantive decisions to be made at the local level, "giving on-site inspectors the authority and discretion they need to be more than just policemen," encouraging them to be consultants to the industries they regulate in addition to being enforcement agents for the agencies they represent.[26]

If, however, the inspector lacks this flexibility, and:

> the regulatory program is organized along such rigid lines that the inspector has no room to maneuver and no authority to decide how a particular situation should be handled, then the skills of the astute, sensitive and innovative inspector are going to be wasted.[27]

Enhancing the inspector's role as consultant can cause problems, especially if his discretion is increased and he is not well trained or his actions can inappropriately bind the agency. But there is little question that acting as a consultant can often increase the inspector's morale.

II. PERFORMANCE EVALUATION

Monitoring Performance

If the worker is bored or frustrated, he is unlikely to perform up to his potential. Similarly, if good behavior is not rewarded, nor poor performance punished, the employee may lack the incentive to work as well as he is able. Before dispensing either penalties or rewards, it is first necessary to assess performance.

Attempting to evaluate inspectors encounters the same type of problems as trying to evaluate the behavior of inspectees. It may, of course, be simpler for the agency to assess the inspector. Inspectors are often fewer in number and perform less varied and less complex activities. Moreover, the agency may have more power to control the inspector's actions, particularly if he is an agency employee. Yet for both inspector and inspectee, the agency may have difficulty in operationally defining good behavior as well as in monitoring for it.

Whatever criteria are used to define good performance will affect the inspector's incentives and conduct. If a housing inspector is rated by the number of complaints handled, he may "increase productivity" by inspecting more premises more quickly, doing a less thorough job on each one.[28] Focusing on the easily measurable items means less emphasis may be placed on the intangible yet possibly more important aspects of performance.

While the agency cannot completely monitor inspector behavior, there are various approaches that may be used to estimate on-the-job performance. One method is to require the inspector to keep records of his activities. These documents provide only some evidence that the job was done, even less indication about how well it was performed. Moreover, while additional paperwork may enhance accountability, it tends to decrease work satisfaction. Here, as elsewhere, there may be trade-offs between performance monitoring and job quality.

A second approach used to evaluate inspectors is to oversee their activities directly. This is less costly when many inspectors work in close proximity. It is not surprising that inspectors in FDA laboratories receive more intensive scrutiny than their counterparts in the field.

As the inspectee's conduct changes when the inspector is present, so too will the inspector alter his behavior while being observed by a supervisor. (Similarly, the inspectors probably worked harder when we were with them.) OSHA found that the average number of violations cited by state inspectors quintupled when they were accompanied by federal supervisors.[29]

To avoid such bias, the inspector may be watched surreptitiously. Hidden microphones and long-range photographs have been used to detect and prove that building inspectors were extorting money.[30] City investigators have also posed as landlords and contractors to help determine the competency and honesty of these inspectors.[31]

Another way to monitor inspector performance is to make reinspections. But a reinspection can be costly, not only to agency, but also to the inspectee. Moreover, a reinspection may result in a different outcome than the original, not due to inspector error, but because the conditions changed. And inspectee behavior can often be rapidly modified. Thus reinspections usually furnish more relevant evidence about the quality of the initial inspection when products or equipment are being examined. It makes more sense to use reinspections as a device for checking the performance of an elevator inspector than it does for a restaurant inspector or an OSHA compliance officer.

An additional method of evaluating the inspector is to have him examine products of known quality. "Test" cars have been taken to motor vehicle inspection stations to discover the accuracy of inspectors under normal working conditions. Meat inspection supervisors historically placed portions of trichinosed muscle into specific tins to determine if they would be detected.[32]

The monitoring of inspector performance can be aided by complainers, informers and other inspectors. Complaints provide only a limited source of information. Elevator owners, for example, are usually not knowledgeable enough to complain. Many inspectees may not complain for a variety of reasons, including fear of retaliation. Builders who refuse to pay extortion are targets for harassment by corrupt inspectors and other government officials on-the-take.

The information that does come from complaints may be self-interested and biased, but it can prove useful in pinpointing types of problems and particular inspectors who bear watching. The FAA receives the views of manufacturers and carriers concerning individual inspectors. If the inspector is disliked, his work will be more closely observed. Interestingly, the same holds true if he is praised too highly.[33]

Informers sometimes provide helpful information. In the close confines of the fresh fruit and vegetable market, for example, favoritism toward a particular supplier will quickly be noticed by the others. In elevator inspection, incompetence will rapidly be detected by the repair team that is present throughout the examination.

An inspector usually knows the quality of his colleagues' work, as well as whether they are receiving illegal gratuities. But he is unlikely to tell, particularly if the information is derogatory. This is especially true if retribution may follow. Information about corruption comes primarily from outcasts or turncoats. The major break in the 1970s New York City building inspection scandals occurred when an inspector was induced to tape-record office conversations in exchange for immunity.[34]

Inspectors are more able to monitor each other's work when they are in teams, especially rotating teams. But they still may be reluctant to provide the agency with useful information. An effective agency strategy is to have one inspector responsible for work initially performed by another. The Coast Guard uses three "layers" of examination by different inspectors, who are not co-workers. This leads to some duplication, but it also provides built-in checks. There are inspectors located at the manufacturing plants who inspect ship components. Then there are resident inspectors in shipyards who examine ongoing construction. And finally there are separate repair yard inspectors who re-examine the vessel. Factory-inspected items are re-inspected during construction and installation. And new ships are re-inspected, often at a different port, within months after going into service. The possibility of catching an inspector who approves substandard work is thus quite high.[35]

Most agencies have neither the opportunity nor the resources for such layered inspections. They use one or more of the other methods of evaluation and usually receive a general impression, but rarely precise information, about inspector performance. It is simply harder to judge the work of an inspector compared to that of an assembly-line employee. First of all, it is usually more difficult to find objective criteria to accurately gauge his performance. More important, since an inspector's work is usually in the field, it is much harder to monitor his activities. (New York City, for example, recently discovered that many of its fire inspectors were not doing their work. Some actually held other jobs. The establishments they claimed to have inspected included vacant lots, burned-out buildings and non-existent addresses. The vast majority of merchants interviewed stated that their premises had not been inspected as alleged on the official worksheets.)[36] We asked each inspector how he thought he was actually evaluated. A typical response was "I wish I knew."

Punishments and Rewards

To provide the inspector with the correct incentives, it is necessary, but not sufficient, to know how well he performs. He must also be properly rewarded or punished. There are a variety of positive reinforcements: praise, esteem, educational opportunities, pay and promotion. Pay and promotion are probably the key rewards, for the others are highly correlated with these. Penalties include not promoting, not giving raises, suspending and firing.

An agency has less control over an inspector who is not an employee. This is particularly true if inspection is not the worker's primary occupation. The Massachusetts government, for example, has little influence over the private garage and gas station attendants who perform the mandatory semi-annual motor vehicle inspections. The state does not control either the worker's pay or promotions. This is one of the reasons why the quality of the automobile inspections is so suspect.

The typical inspector is a government employee working under civil service regulations. Civil service rules emphasize job security, which can enhance perceived job quality and consequently performance. Normally, however, by reducing or eliminating the differential between the payoffs for good and bad performance, civil service regulations decrease the incentive to do a good job.

It is more difficult for government administrators to penalize their employees for poor work than it is for their private sector counterparts. While one can refuse to promote workers for poor performance, the more stringent penalties—demotion or firing—can rarely be imposed. This is one reason for requiring a probationary period for new employees. Civil servants can, of course, be dismissed, fined or imprisoned for outright corruption. For effective incentives, however, it is important that the punishment fit the crime.

Inspection agencies have been accused of being both overly lenient and overly harsh in their treatment of wrongdoing. A New York Times report described the INS as "an agency eager to keep its misdeeds hidden, and when it cannot, reluctant to administer more than token punishment to wrong-doers."[37] While former immigration personnel claim to have witnessed many beatings, 1979 marked the first time that the federal government brought brutality charges against INS agents.[38]

By contrast, the USDA stands accused of not sufficiently supporting its inspectors. A Harper's article asserted that "every inspector has dozens of anecdotes about the failure of USDA supervisors to back him up in disputes with plant management."[39] When forty Boston inspectors were accused of corruption in 1971, union officials claimed that the agency decided "to throw these men to the wolves."[40] On the other hand, industry personnel who had offered bribes were granted immunity. GAO investigators reported low morale among inspectors, in large part due to the general failure of USDA to support them.

Inspectors have a number of career paths. One is an upward progression within the inspection field. In the Coast Guard, junior officers can rise within the inspection area from Lt.J.G. through full lieutenant, commander and captain. The top of the career ladder is admiral in the Office of Marine Safety.[41] A second career path leads out of inspection into broader managerial or professional responsibilities. At the FAA, they try to permit flexibility of choice by internally providing the opportunity for employees to cross from one career area to another.[42]

For state employees, the possibilities for advancement are more limited. Indeed, a cited reason for expecting good performance by the state surface mining control inspectors was the chance, after 1977, to jump to the more attractive federal inspection jobs.[43] This opportunity, however, remained open for only a short period of time.[44] As is the case with many types of inspection, the principal career path led to industry, raising the spectre of capture.

Inspectors with low work satisfaction, whose job quality is poor and job usefulness uncertain, inspectors whose performance is not capably monitored by the agency or whose agency does not relate rewards and punishments to performance—

these are inspectors with little incentive to do a good job. There are also inspectors who are highly susceptible to the Moral Hazards.

III. MORAL HAZARDS

Capture

Introduction

There are many reasons why an inspector may not do a good job. He may, for example, be spiteful or sick or lazy or incompetent. An inspector does a poor reporting job when he misses important violations, or cites unimportant or nonexistent ones. A bad inspection may thus mean reporting too many as well as too few infractions. If the inspector is captured, he will cite too few violations, especially too few serious violations, because of empathy with or pressure by the inspectee.

The primary beneficiaries of the regulations, and thus the inspections, are generally diffuse and unorganized interests— elevator riders, pedestrians, apple eaters, swimmers, nursing home residents, etc. Many of these never see the inspector, nor understand his work. Rarely will they reward him for doing a good job, or punish him for a bad one.

By contrast, the inspectee knows the inspector. Face-to-face meetings are common; often there is continuing interaction. The inspector has the power to impose large costs on the inspectee. The inspectee thus has the opportunity, the motive— and often the ability—to influence the inspector's behavior.

The inspector's job entails finding fault with the inspectee. While the inspector may occasionally help the inspectee—perhaps by pointing out overlooked problems or suggesting ways to increase efficiency—his primary function is to discover and cite deficiencies. Such work is stressful. It is not surprising that the inspector would attempt to lessen the stress. One method is to attempt to "capture" the inspectee, making him sympathetic with the inspector's plight, getting him to realize that the inspector is only doing his job. This winning-over of the inspectee is usually socially desirable. Another way for the inspector to decrease stress is to become captured. He may come to understand the problems of the inspectee, convince himself of the inspectee's good intentions and the general merits of his case.

The inspector and the inspectee can impose costs on each other. The inspector can come at inconvenient times, can disrupt activities, can antagonize personnel. The inspectee can withold information, complain to the inspector's supervisors, and generally make his task difficult, his job miserable. The inspector and inspectee are both better off if such costs are kept to a minimum. Society also benefits—so long as the inspector does a good job. Keeping the encounters amicable can increase inspectee cooperation, making it easier to discover problems and more likely that they will be readily rectified. The danger is in the cooptation of the inspector, his becoming overly conciliatory and complacent.

The inspectee may agree to impose few costs on the inspector, and indeed to reward him, in exchange for the inspector being less strict than is socially desirable. The inspectee can reward the inspector with money and future job opportunities, but perhaps more important, with respect and even friendship. Inspectors, like most people, seek approval. And they are more likely to receive approval by attaining a community of interests with the people they regulate than they are with the uninformed, unorganized and diffuse public.

Capture of inspectors is similar to, but not identical with, the capture of the regulatory agency. Top agency personnel are more prone to be captured by the industry as a whole. Even captured officials may take actions that injure particular firms, if most of the companies benefit. Individual inspectors, on the other hand, do not see trade association or industry representatives. They generally examine inspectees individually, and receive pressure from them individually. Being lax in the finding and citing of infractions for each inspectee could actually injure the industry as a whole. In other words, the regulated firms might be better off if there were no inspector capture. This could be the case where the regulations provide some benefit for the entire group, such as weights-and-measures, motor vehicle and grain inspections, and perhaps even for nuclear power plant and nursing home regulations. Certainly the capture by students (inspectees) of professors (graders and recommenders) may not enhance the position of students as a whole.

Factors Affecting Capture

Capture is not an either/or proposition. Inspectors may be more captured or less captured, or lax about reporting different types of violations, or remiss at different times. Nonetheless, for simplicity, we will discuss capture as if it were a simple, homogeneous state.

Our focus is on the ability of the inspectee to capture the inspector. The inspectee, however, does not always have the incentive. Capture is superfluous if the inspectee has nothing to hide, and the inspector is fair and competent. Capture is unnecessary if the inspector is fair but incompetent, incapable of discovering violations. There is also little need for capture if enforcement is ineffectual. For example, since serious penalties are rarely imposed, landlords do not expend many resources trying to win over the housing inspector. This is one explanation of why so few housing inspectors appear to be on-the-take.

Generally, of course, inspectees would like to capture the inspector. They want to be assured, at minimum, of a fair and impartial hearing. Most would prefer a sympathetic one. They want the inspector to keep inconvenience costs as low as possible. Few inspectors are incompetent. And most inspectees are violating some regulations. In many areas of inspection there are enough strict standards so that every inspectee is committing some infractions.

Capture is thus usually beneficial to the inspectee. Agency policy can affect the size of these benefits. It may not pay, for example, to invest resources to capture a resident inspector if his stay will be only a brief one. This is one reason why agencies rotate inspectors.

The ability to capture the inspector depends on the characteristics of the inspector as well as on the opportunities and power afforded to the inspectee by the inspection environment.

1. Inspector Characteristics.   The background of the inspector affects the difficulty of capture. Those with backgrounds similar to the inspectee are more prone to identify with his problems, to be sympathetic with his plight. Judges, for example, are more lenient with white-collar criminals and white-collar crimes.[45] That "30% of U.S. Rail Safety Inspectors Monitor their Former Employers" is newsworthy because of the implication that this makes them more susceptible to capture.[46] The top-level PSRO "inspectors" who examine and evaluate physician behavior are also doctors. One author argues that there is no need for physicians to capture these inspectors. They are already identical to them.[47]

If the inspector is a professional, especially with training different from the inspectee, he is less likely to be captured. The profession provides identity, support and approval for the inspector. Nursing home surveyors are usually registered nurses; animal health inspectors are often veterinarians; Coast Guard inspectors are military officers. The nursing home surveyor is a nurse first, an inspector second. A nurse's training and work experience are directed toward providing proper care for patients. To capture a nurse-surveyor requires overcoming his professional background and ethics.

There are other reasons why professionals are less easily captured. It is usually more costly for them to be caught taking favors or bribes. They have a great deal to lose—their professional standing. Many professionals have attractive employment opportunities; their future career is not necessarily controlled by the regulated industry.

A particular background or training, of course, does not eliminate the possibility of capture. Inspection is still stressful, and pressures are applied by the inspectee. ASME (American Society of Mechanical Engineers) Boiler and Pressure Vessel inspectors, for example, are professional engineers. Yet for them, as for most inspectors, "alleged violations are not always a happy situation for either the manufacturer or the commissioned inspector involved."[48] It would be easier for both if there were no violations or, at least in the short run, if no violations were detected.

2. Inspection Environment. Inspectors can be captured by persuasion as well as by tangible rewards and penalties. Capture is most complete if the inspectee can win over the "hearts and minds" of the inspectors. This is most easily accomplished if there is frequent personal contact, allowing for discussion, argument and explanation as well as flattery, proferred friendship and other forms of psychic inducements designed to create sympathy for, and hopefully empathy with, the inspectee's problems. The more tangible rewards the inspectee can offer include monetary bribes and promises of future employment. The penalties include being uncooperative, filing complaints about the inspector, and threatening him with physical harm.

a. Psychic inducements. Inspectors are more susceptible to capture the more personal contact they have with the inspectee, and the less contact they have with those of differing views. The inspector is more likely to become sympathetic with the inspectee when they meet face-to-face, when examinations are long, and when there are repeat encounters. He is more easily co-opted when he monitors only one industry, and most particularly, only one firm. And he is most easily swayed to the inspectee's viewpoint when he meets with no one else—neither the beneficiaries of the regulation nor knowledgeable third parties.

The meat and poultry inspector works in an environment that is highly conductive to capture. While MPIP employs thousands of resident inspectors, there are so many small processing plants that most of these inspectors work alone. Except in major cities, they are not generally rotated. The typical inspector thus works all day, every day monitoring a single firm.[49] As one inspector put it:

[We] go out among the regulated to do our job. We don't just visit them periodically, we just about marry them. Day after day, night after night, we are in the lion's den alone with the lion. How are we supposed to get along? USDA doesn't tell us. How are we supposed to resist the barrage of threats and temptations the packers constantly direct at us? USDA doesn't tell us. USDA does tell us to use our ingenuity to do our job, to use our common sense—but that's not very helpful when you're in the lion's den.[50]

A survey of twenty resident inspection programs concluded

that MPIP "has the greatest potential capture problem of all programs surveyed."[51]

The nature of the U.S. grain inspection system also facilitated capture. Most of the federally licensed agencies that inspected export shipments, for example, did so at one or perhaps two elevators. Some individual inspectors monitored a single elevator for over 15 years. Not surprisingly, a U.S. attorney argued that:

> The fault . . . throughout the system. . . is in the intimate relationships, the mutuality of interests, that has developed between the elevator companies and the inspection agencies, where the personnel of the inspection agencies, in effect, feel that they are servicing the elevator. . . . They feel that their loyalty is to the elevator. Many of them show a down-right open hostility toward the U.S. Department of Agriculture.[52]

A GAO report on nuclear power plant inspection concluded that some loss of objectivity was inevitable the longer a resident inspector remained at the site.[53]

> While on site, the resident inspector will probably become acquainted with many, if not a majority, of the plant employees on a first-name basis. The longer that the resident inspector remains at the plant, the more he may consider himself a part of that plant's organizational structure. He may even begin to defend the plant against outsiders who raise questions about plant activities. The resident inspector may tend to regard such questions as a reflection on his performance and professional judgment.[54]

It is not only on-site activities that can result in loss of objectivity. Fraternization can occur off as well as on the job. Because of the danger of capture, agencies write directives specifying codes of conduct for inspectors. The NRC, for example, prohibits the resident inspectors from socializing with plant employees in any way, such as having lunch or carpooling to work with them. The inspector is also supposed to guard against his family becoming too involved with the families of plant workers.

He must worry about what clubs his spouse joins, and what friends his children make. The actual enforcement of such rules is, of course, difficult—especially in isolated areas where nuclear plants are often located.[55]

Resident inspectors most easily lose their objectivity. Rotating inspectors, and having them work in teams can reduce this danger. But such tactics are not without problems. Rotating inspectors, especially in rural areas, could mean forcing them to move their homes. This is costly, not only to the agency, but also to inspector morale.[56] Moreover, while rotation may lessen the likelihood of capture, it can also mean the loss of useful plant-specific information. It certainly increases the difficulty of building alliances with interested third parties, as well as with individuals and groups inside the inspected firm. This decreases the inspector's ability to detect violations.

Although limiting the danger of cooptation, rotation can decrease the likelihood of inspectee cooperation. Important elements in achieving cooperation, especially when there is continuous monitoring, are constant communication, uniform enforcement, and a "personal chemistry" that often can only be built up over time.[57] A meat slaughterhouse manager explained what he liked about the MPIP inspections:

> There are no surprises at our plant. The inspectors never find something that you didn't know existed. They keep us informed, and we always know what is happening and why.[58]

Rotation increases the possibility of "surprises."

Teams help prevent the loss of objectivity by providing support for each inspector. They also lessen the possibility of capture through bribery—by increasing the danger of exposure; and through intimidation—since there is strength in numbers. But teams can be a costly and inefficient way to monitor activity. To use teams, for example, MPIP would have to almost double the number of resident inspectors at the processing plants. Neither rotation nor teams provided the answer for the FAA. After it discovered it could not effectively supervise its employees or otherwise decrease the danger of capture, it discontinued its resident inspection program.[59]

It is not only resident inspectors who can lose objectivity. Restaurant inspectors are sometimes captured.[60] They interact almost exclusively with restaurant management—often over coffee. While they are continually "rotating" among local restaurants, they visit establishments often enough to soon develop an informal relationship with many restaurateurs.

Nursing home surveyors, on the other hand, while their examinations take longer, should be less prone to lose their objectivity. Not only are they professionals, but they also work in teams. Perhaps most important, they talk to the beneficiaries of the regulations—the patients—as well as to knowledgeable and concerned third parties, such as social workers and physicians.

EPA and OSHA inspectors are less likely to be captured because they deal with so many industries, each of which has somewhat different interests. Moreover, they may never get to know any inspectee intimately.[61] OSHA inspectors should rarely lose objectivity since, especially in unionized plants, they also meet with the workers whose safety and health they are trying to protect. One inspector told of an inspection in which he was involved where there were ten subcontractors and ten unions. An entourage of twenty people accompanied him on his rounds.[62]

While monitoring many different kinds of firms can decrease the likelihood of capture, it can also lessen the inspector's ability to do a good job. The President's Interagency Task Force of Workplace Safety and Health concluded that OSHA inspectors

> know too little about the industries and operations they are inspecting. . . . Inspectors who are required to be jacks-of-all-trades . . . cannot develop the familiarity with particular worksites and industrial operations that would permit confident site-specific evaluations.[63]

As economists love to point out, there are trade-offs in this world. Resident inspectors, for example, may be most able to detect violations, but they are also most prone to capture. Methods to decrease the likelihood of cooptation may at the same time lessen the inspectee cooperation which is often essential for effective regulation. Reducing the danger of capture could also decrease the inspector's reporting ability.

b. Tangible rewards. Inspectors can be captured by the lure of tangible rewards. These include favors and monetary bribes, future employment, and, for some inspection agencies, sales.

Many of the factors that endanger objectivity also increase the possibility of successful bribery. Face-to-face contact simplifies payment. So does continual interaction. The absence to beneficiaries, third parties and, often, other inspectors, decreases the danger of exposure. Resident inspectors are particularly likely to be captured, not only by psychic inducements, but also by material rewards.

Scandals rocked both the meat and grain inspection systems in the 1970s. Meat inspectors commit a felony if they receive *anything* of value from the regulated industry. Indeed, when the initial indictments against forty Boston meat inspectors were handed down in 1971, they included such minor charges as receiving a handful of screws, a light bulb, a half can of shoe polish, and a ride home (for the inspector's daughter). The meat inspectors argued that accepting such small gratuities actually improved their ability to do a good job. To decline a gift creates ill-will.[64] To have refused a bag of doughnuts for the night shift, or a Thanksgiving turkey or the Christmas scotch—items routinely given to plant employees—would have made plant management edgy, suspicious and less cooperative.[65]

From the inspector's viewpoint, accepting meat was even less suspect. The packers threw away hundreds of pounds of edible product daily. An inspector could observe police and firemen and politicians, veterans groups and hospitals, as well as the packer's employees, leaving the plant laden with free meat. It was hard for the inspector to understand why he alone should refuse the proffered gift. Where inspectors worked together, new ones immediately felt pressure to accept this "cumshaw":

> We are weaned on the tradition. The old-timers always say 'It isn't a good inspector who pays for his Sunday dinner.' They tell you that everybody else does it and has always done it, that

it has nothing to do with doing your duty, and that if you don't take it, someone else will. I figure the job is hard enough without having the other inspectors suspicious of you.[66]

It is, however, difficult to defend accepting hard cash from the inspectee. In 1976, thirty-one New York meat inspectors were indicted, some for receiving $20–$75 weekly from packers and processors. Nor surprisingly, the meatpackers argued that these were not bribes to capture the inspectors but instead were extortion payments to avoid harassment.[67]

The export grain scandals of the mid-1970s also exposed a system of favor and bribes. Inspectors were sometimes entertained by elevator personnel with expense-account lunches. "And, of course," testified one former inspector, "we all got turkeys at Christmas." Other witnesses reported more serious infractions. One, for example, received payment from the grain company for five hours a week of unearned overtime, quarterly bonuses of $400 and a Christmas bonus of $1,000.[68]

Resident inspectors are not the only ones who can receive gratuities. Inspectors of restaurants, for example, have also accepted bribes. When undercover agents, posing as health inspectors, visited twenty randomly chosen New York City restaurants, they were offered cash bribes by four owners; three others invited them to enjoy a free meal.[69] Restaurant inspectors sometimes have the power to extort such payments. The power to extort is discussed in the next section.

Future jobs are also a reward for some inspectors. If inspectors plan to seek employment with the regulated industry, it behooves them to maintain a friendly rather than an adversarial relationship. Nuclear engineers, for example, who are among the most highly skilled and highly paid inspectors, often go to work for the nuclear industry. Many American Gas Association inspectors find employment with the firms they evaluate. And strip mine inspectors frequently become mine employees.[70]

In a few areas, inspectees can reward their monitors not only with money and possibly jobs, but also with inspection business. This can occur if the regulated are allowed to select the inspector. Many may then flock to the inspectors who are lax. If reimbursement is based on the number of inspections—as is the case when the inspectee pays directly for it—competition among inspectors may lead to a deterioration of inspection quality.

Grain shippers can sometimes "shop for grade." Warehouse-men's samples, for example, can be submitted to any official inspection agency. This places the inspectors in a competitive position, and has lead some agencies to complain that others take away their business by grading too leniently.[71] The problems created, however, are not always severe since domestic grain buyers are generally knowledgeable and continually make purchases. If grades are inaccurate they can complain to the authorities, refuse to use grading information for subsequent purchases or, most threatening, take their business elsewhere. Knowledgeable grain buyers make this voluntary, domestic grading system somewhat self-policing.[72]

Since no product buyer is involved, motor vehicle inspections lack this potentially self-policing mechanism. Although virtually all automobile inspectors have the ability to carry out a competent inspection, in many states they do not have the incentive to do so. In Massachusetts, for example, there are over 3000 inspection stations. The motorist can go to any one, and for a small fee, receive the mandatory automobile inspection. Most car owners seem to want a quick, superficial and passing inspection, and this is what they get. Many stations, for example, are loathe to fail a regular customer whose purchases provide their main business. They are captured.

Motor vehicle inspection is interesting because there exists an opposing, although usually a weaker incentive. Since inspection stations often do repair work, it may pay them to cite violations in order to increase their parts and repair business. Although motorists can get repairs anywhere they like (or even try to get a passing sticker from a more lenient inspector), many purchase their repairs from their inspector. This gives some inspectors an incentive to perform a tough inspection—or since car owners are often not knowledgeable, at least to claim to have found a lot of violations.

c. Tangible penalties.   Inspectees can wield the stick as well as offer the carrot. They can be uncooperative, withhold information, and generally make the inspector's job more difficult. They can complain to the agency about the inspector.

Indeed they are the principal source of grievances against him. They can put social pressure on him. And finally, they can turn ugly. They can verbally abuse him, or threaten him with physical injury, or actually do him harm.

While few inspectors in the field are immune from some sorts of negative pressures and intimidation, it is resident inspectors who are most easily threatened. USDA files, for example, are filled with instances of verbal and physical attacks on the resident meat and poultry inspectors.[74] The Nuclear Regulatory Commission relies very heavily on the contractors' own quality control inspectors in monitoring construction. These inspectors are continually on the scene, but since they are working for the regulated utility, they have little actual independence. To some extent they are already captured. A former quality control inspector at the South Texas Nuclear Project reported that the construction management lectured inspectors, not about the importance of insuring quality, but about their role in controlling costs and preventing delays. When the inspectors continued to point out construction flaws, this former quality control expert testified, his life was threatened by construction workers and a colleague was beaten up. For five months the inspectors were afraid to inspect construction and thus failed to note, among other things, bad welds and deficient concrete pouring. They simply falsified reports.[75]

Federal supervisors have also been intimidated. USDA grain investigators, who monitored the private, licensed inspectors, sometimes worked in fear. The automobile tires of the supervisors were slashed. They were "jostled" while in elevators cages high above the ground. One company described its waterfront loaders as "tough guys [who] don't like those creeps from the Department of Agriculture always looking over their shoulders." There were even rumors of a contract to break the knee caps of a supervisor who had reported irregularities.[76]

Face-to-face contact is not even essential for intimidation. Recently, in Massachusetts, a state auditor who had exposed MBTA misdealing, quit his job after receiving numerous written and late-night telephone death threats.[77]

Virtually all field inspectors are in some danger of harassment. One inspection area where intimidation should be rare is motor vehicle safety, particularly in states where large numbers of private firms perform the examinations. In Massachusetts, for example, if one inspector is too tough, the motorist merely has to find another one among his 3000 options who isn't. Yet even here there is no assurance of safety.

In May 1981 a service station manager tried to place a rejection sticker on the windshield of a vehicle that had failed inspection. The car owner jumped out of the car and began chasing the manager around the lot. The dispute escalated rapidly. The motorist got back in his car and allegedly bumped the manager with his vehicle. Whereupon the manager grabbed a hammer and smashed the customer's windshield. The irate motorist sped off, but returned shortly with a companion and two baseball bats. "That's when I called the police," the manager said.[78]

Extortion

This book distinguishes between capture and extortion. From the inspector's perspective, capture is more passive, extortion more active. The inspector extorts; he becomes captured. In capture, it is generally the inspectee who initiates, who tries via persuasion and psychological inducements, or material rewards, or tangible punishments, to affect the inspector's behavior. The inspector, of course, may be receptive to the positive overtures. In extortion, by contrast, the inspector is the aggressive party, demanding or soliciting tribute to act in a prescribed manner.

If the inspector is captured, it means he will be lenient; he will be lax in the reporting of infractions. He will not be doing as tough or as good a job as is socially desirable. In pure extortion, on the other hand, the inspector requests an illegal payment to perform competently. The inspectee pays a bribe in order to receive fair, not necessarily favorable treatment. He pays to get the inspector to do his job—e.g., to arrive on time— and to prevent harassment, such as the citing of picayune or even nonexistent violations. In extortion, the inspector (ab)uses his official power to apply improper pressure on the inspectee in order to receive tangible rewards. The extortionist does not use persuasion or psychological inducements to get his way. He threatens punishment.

Bribes can sometimes be a combination of both extortion and capture. The inspector demands a bribe in order to act fairly. The inspectee is willing to pay even more to receive lenient treatment. In the real world, when bribes are paid, it may be difficult to demonstrate that it was extortion, attempted capture, or both. The law does not usually distinguish between them.

Extortion is much rarer than capture. The inspector-inspectee relationship is already adversarial, and extortion increases that conflict. Capture, by contrast—at least capture by positive means—lessens the antagonism. More important, in extortion, the inspectee is a definite loser. He has an incentive to complain, to try to eliminate the pressure and punish the extortionist. By contrast, capture often consists of a voluntary exchange in which both parties benefit. It is only society that is made worse off.

Consider the case of a corrupt meat inspector, willing to misuse his authority to enhance his personal income. He might try to extort the packer, threatening that unless payment were forthcoming, he would cite good meat as defective. Alternatively, he might indicate his willingness to be captured—to cite borderline meat as acceptable in exchange for a bribe. Clearly the inspectee would prefer the latter situation. He may well find the transaction a profitable one. He is more inclined not only to pay the bribe, but to proffer his goodwill. He is less likely to try to intimidate the inspector, to complain about him, or in any way to publicize the illegal exchange. The inspector is thus more likely to receive his bribe, and less likely to get caught. Rather than proclaiming "I am an extortionist," the corrupt meat inspector is much better off signaling "I can be captured."

Successful extortion requires:(a) the power to make a credible threat; and (b) the ability to escape punishment. Many, if not most, inspectors lack these requisites. A Massachusetts motor vehicle inspector, for example, cannot extort money from the automobile owner. He does not have the power. Suppose he threatened not to certify a satisfactory automobile unless he were given a bribe. Why should the car owner pay? There are thousands of other independent inspection stations available. Many of these would value the motorist's inspection business and his goodwill. There is capture in this market, but not extortion.

The typical housing inspector also has little leverage. He might, for example, threaten to cite nonexistent violations until he is paid off. But why should the landlord pay anything? Even if taken to court, the judge would probably accept as sufficient the owner's promise to repair the (nonexistent) problem. And finally, a different inspector could eventually verify that the reported violations were no longer in evidence.

The motor vehicle inspector lacks monopoly power. The housing codes are not vigorously enforced. Neither inspector, moreover, can prevent the irate inspectee from complaining to higher authorities, possibly leading to reprimands, suspension, loss of job and pension, and even to jail.

Some inspection environments are conducive to extortion. Through abuse of his authority, the inspector can impose significant costs on the regulated—enough so that the inspectee is willing to pay a bribe. Additionally, the inspector has protection from exposure. The threat and payment are not witnessed by outside parties, and the victim himself does not desire that the extortion be detected—either before or after payment.

One reason an inspectee might keep silent is that he remains under the inspector's power. Even after paying thousands of dollars in extortion money to liquor license boards, for example, few restaurateurs are eager to complain. Retaliation can be swift and sure. Restaurants, with their fragile clientele, are easy prey for official harassment. And excuses can usually be found for license revocation.

A second reason not to complain is if the inspectee believes that a kind of "capture" was involved—that the inspector has been lenient with him. Corrupt policemen, for example, have little difficulty extorting payments from conspicuous but illegal gambling operations. Conspicuousness makes it easy for police to arrest or otherwise harass employees and players of non-paying establishments. Illegality means the gamblers cannot complain. A numbers operator at the New York Knapp Commission hearings in 1970 explained:

> You can't work numbers in Harlem unless you pay. If you don't pay, you go to jail .... You go to jail on a frame if you don't pay.[79]

An environment conducive to inspector extortion is new construction. Here the inspector has great power. His approval is necessary before construction can begin, before various phases of it can continue, and most important, before the building can be occupied. Delays can cost the contractor tens of thousands of dollars per day. And the inspector can create delays not only by finding problems, but also by simply not showing up on time.

Extorted contractors usually believe they had better pay. In the 1960s, one builder stated, "I suppose it's possible to do a profitable business in Chicago without greasing anyone's palm. But it wouldn't be easy, and I don't know of anyone who's willing to try."[80] Most corrupt inspectors do not directly ask for bribes, "but you know what they're hinting at. If you play dumb and don't come across, they start showing up a week late."[81] A New York contractor testified that if you did not pay off the inspector, it could take two months to receive final approval for electrical work. How long would approval take if you met the inspector on the job and gave him some money? "You get it right away."[82]

Building codes are usually so complex, with standards so rigid and outdated, that every construction project violates many of the regulations. This gives even the most punctual inspector a legal (if often inappropriate) authority to hold up any construction process. The inflexible and outmoded standards increase the building inspector's extortionary power. As one New York City contractor explained, with that code book, "the inspector has a gun in his hand."[83]

In 1961 the *Wall Street Journal* asserted:

> The most common building material in this city is neither steel nor concrete, brick nor wood. It is graft.[84]

Thirteen years later, a formal investigation of New York City building trades reached the same conclusion. Payoffs to inspectors were common:

> We found that corruption and its cover-up are endemic to the working day of many building inspectors. Throughout their day

they contrive schemes through which they can be bribed and
methods for covering up the bribes and gratuities they receive.[85]

The contractor pays the extortion money to avoid costly delays. Even after receiving the final certificate of occupancy, it
is usually foolish for him to report the bribe. This is especially
true if the contractor believes the system is corrupt. By revealing
that he made illegal payments, he incriminates himself. More
important, he is still vulnerable, for like all construction projects, his is legally "substandard." Citations against his building
can, at minimum, hurt his reputation. Additionally, all his subsequent work in the area is subject to inspector approval. Colleagues of the accused inspectors thus have the power to retaliate
effectively against him.

Reporting extortion can create huge problems for the contractor. The ultimate benefit, if any, is a public one—a less
corrupt inspection system. The contractor who fights city hall
may promote social welfare; he himself is almost certain to be
a big loser.

This discussion is not meant to imply that all building inspectors extort, or that there is some corruption in all building
inspection departments. It is just that the new-construction inspection environment is more conducive to extortion than most.
The principal basis for the inspector's extortionary power is
that his approval is necessary before operations can continue
and before occupancy is allowed. The fact that no building is
ever in strict compliance with the regulations merely increases
that power.

Old buildings are subject to many of the same complex standards as new construction. And building inspectors typically
spend much of their time making safety inspections of existing
structures. But closing a building down is a great deal more
difficult than delaying an opening. Long, drawn-out judicial
proceedings may be required, and during that time the structure can remain in use. If the requisite corrections are finally
made, the building will never be closed. Time is thus on the
side of the owner. Procrastination by the inspector imposes
few costs. In this situation, extortion requires the threat of
clear, often conspicuous, and illegal harassment. Not sur-

prisingly, extortion is not a usual feature of this inspection process.

An inspector's approval is necessary before grain can be exported from the United States. Delays are costly. Grain inspectors thus possess extortionary power. Indeed, the corruption in the Gulf Coast export grain inspection system was exposed when a ship captain balked at paying an extortionary demand he considered excessive. (The owner offered $2,500; the inspector wanted twice that amount.)[86] Grain inspectors have power to extort ship captains; the inspectors are more likely to be captured by the grain companies. This is especially true for resident examiners, continually interacting with the same elevator workers and management, and thus vulnerable to their pressures. The grain companies have the ability to punish resident inspectors, and in particular, to meet extortion attempts with credible counter-threats. The bribes paid by the grain companies almost always bought lenient inspections and generous grades.

Prior approval by an immigration inspector is required for lawful entry into the United States. This fact gives the inspector some power to extort. Extortion of legal arrivals is, however, extremely rare; the inspectee has both the incentive to complain to higher authorities, and often the ability to gain satisfaction. Immigration officials can, however, successfully extort illegal aliens. This is often a combination of extortion and capture. The aliens usually cannot complain without admitting their unlawful presence in the United States.

Since illegal aliens generally have little money, extortion demands are sometimes for sexual favors. The provision of border-crossing cards and entry permits to female aliens, including prostitutes, in exchange for sexual relations has been reported by many ex-officials. It is not only women crossing the border who are vulnerable to sexual extortion. It has also been claimed that Border Patrol agents arrest alien women on inspection tours of Texas towns and demand sex in return for their release.[87]

A great deal of the behavior of the Border Patrol is unsupervised. Their actions and decisions cannot usually be rechecked. The job is often dull, occasionally dangerous, and usually frustrating. The Mexican border has become a sort of no-man's-land, where an official can become his own law. There is enough violence on both sides to make brutality come easily. This is especially the case since aliens are often regarded as wetbacks and tonks, and because confessions are sometimes the only way to prove unlawful entry. Moreover, beatings, rapes and extortion by the Border Patrol are difficult to prove since they usually consist of the testimony of an alien versus that of a federal official.[88] This is, in other words, an inspection environment more conducive than most not only to the moral hazards, but also to physical abuses of power.

IV. SUMMARY

To do a good job requires not only the ability, but also the incentive. Any employee will perform better if he finds his work satisfying—if, for example, the work is inherently interesting, he has some autonomy, and he is able to use his skills and training. The work will also be more fulfilling if it seems useful—if the goals of the enterprise are laudatory ones and the employee believes he contributes toward their achievement.

In any occupation, the employee will also try harder when good work is appreciated and rewarded, while careless or shoddy work is appropriately punished. This requires the employer to be able both to monitor performance and to create and implement a fair reward system. Workers who are geographically dispersed and who provide services rather than produce goods, are among the most difficult to monitor. The typical field inspector is such a employee.

What differentiates inspection from most other jobs is its adversarial nature. The inspector's main task is to find and report violations. Inspectees prefer not to have their deficiencies cited. The inspector's work is thus filled with conflict and stress. One way to reduce that stress is to become captured.

Inspectors can be captured by psychic inducements, tangible rewards and tangible penalties. An inspector is more readily captured if his work is not satisfying, and if the agency does not effectively monitor his behavior. He is more susceptible to psychic inducements if he lacks a professional or outside support group. He will more easily empathize with inspectees if his background and training are similar to theirs. If he monitors only one industry or, particularly, one firm, he is more likely to be won over to that perspective.

Inspectees are better able to gain the inspector's sympathy, and to offer rewards or threaten penalties, if there is face-to-face contact, and especially if there is continual interaction. Resident inspectors are thus particularly prone to capture. While resident inspectors may have more ability than nonresidents to perform capably—to spot emerging problems and to cajole inspectees to correct deficiencies—they are less likely to have the incentive to do so.

Bribes can be used both to capture inspectors and to satisfy their extortionary demands. Capture is different from extortion. In capture the inspector is relatively passive; in extortion he is the aggressor. A captured inspector will be lenient in the reporting of infractions. In pure extortion, the inspector threatens to be unfair to the inspectee. He then exacts payment merely for performing his job honestly.

Successful extortion requires the power to make a credible threat, and the ability to escape punishment. The inspector has great power if his approval is required before the regulated party can proceed with operations. Simple procrastination can then impose severe costs on the inspectee. The inspector can escape punishment if outsiders are not privy to the extortion and if the regulated party cannot complain to his superiors or in other ways retaliate against him. Complaints about extortion are unlikely if the inspectee remains under the power of the inspector or his allies, or if the inspectee is also violating the law. The latter situation combines aspects of both extortion and capture.

A particular inspection environment will predispose the inspector to specific behavioral problems. Agency policies can reduce these susceptibilities, but rarely eliminate them. And such policies usually have associated costs. Having team inspections, for example, decreases the likelihood of bribery, but may strain the agency's finances and limit the ability to examine all regulatees. Increasing the direct monitoring of inspector behavior can also reduce the moral hazards. But this may decrease work satisfaction and limit the inspector's ability to achieve the real goals of the regulations. Here, as elsewhere in the world, there are trade-offs.

NOTES AND REFERENCES

1. Current beliefs, evidence and citations on the relationship between work satisfaction and productivity can be found in Donald E. Klinger, Public Personnel Management (New York, Prentice-Hall, 1980), pp.278-284.

2. David Newell, "Drugs, Fruit & Technology are Customs' Varied Quarry," *The New York Times*, July 5, 1981, p.8E.

3. An exception is Paul Danaceau, "Making Inspection Work: Three Case Studies," U.S. Regulatory Council, May 1981.

4. George L. Vinz, "The Compleat Inspector," *FDA Papers*, September 1968, p.4.

5. Michael Goodwin, "Cold Means Feverish Work for City's Boiler Inspectors," *The New York Times*, February 17, 1969, p.24.

6. John M. Crewdson, "Violence, Often Unchecked, Pervades U.S. Border Patrol," *The New York Times*, January 14, 1980, p.D8.

7. Peter Shuck, "The Curious Case of the Indicted Meat Inspectors: Lambs to Slaughter," *Harper's*, September 1972, p.81.

8. John Kenneth Galbraith, *The New Industrial State* (Boston, Houghton Mifflin Company, 1967), Chapter XV.

9. Michael Lipsky, *Street Level Bureaucracy* (New York, Russell Sage Foundation, 1980), Chapter II.

10. Interviews with OSHA inspectors.

11. Paul Strassels and Robert Wool, *All You Need to Know about the IRS* (New York, Random House, 1979), p. 94.

12. *Ibid.*, p.98.

13. *Ibid.*, pp.5,102-103.

14. Vinz, *op. cit.*, p.4.

15. In Maryland, inspectors physically relocated people in HUD projects, only to descend a decade later to condemn buildings, evict tenants. Jackson Diehl, "Tired of Complaining, Tenants Lose Hope," *The Washington Post*, July 23, 1979, p.1.

16. Elizabeth Midgley, "Immigrants: Whose Huddle Masses?" *Atlantic Magazine*, April 1978, pp.20-26.

17. General Accounting Office, "Prospects Dim for Effectively Enforcing Immigration Laws," November 5, 1980, p.8.

18. Crewdson, *op. cit.*

19. *Idem*, "U.S. Immigration Service Hampered by Corruption," *The New York Times*, January 13, 1980, p.46.

20. Phillip Zweig, "Technology Goes to Sea," *The New York Times*, June 3, 1977, p.7.

21. Timothy B. Clark, "What's All the Uproar Over OSHA's 'Nit-Picking' Rules?" *National Journal*, October 7, 1978, p.1595.

22. Paul Danaceau, "Making Inspection Work: Three Cases Studies," U.S. Regulatory Council, May 1981, pp.44,58.

23. *Ibid.*, p.10.

24. *Ibid.*, p. 8.

25. *Ibid.*, pp.64,79,80.

26. *Ibid.*, p.13.

27. *Ibid.*, p. 14.

28. Lipsky, *op. cit.*, Chapters IV,XI.

29. General Accounting Office, "Workplace Inspection Program Weak in Detecting and Correcting Serious Hazards," 1978, p.5.

30. Paul E. Steiger, "New York, Los Angeles Strive to Halt Payoffs to Building Inspectors," *The Wall Street Journal*, December 7, 1966, p.1.

31. Theodore Lyman, Program Director, *Corruption in Land Use and Building Regulation*, SRI International, Vol.1, 1979, p.176.

32. Gerald Leighton and Loudon Douglas, *The Meat Industry and Meat Inspection* (London, Educational Book Co., 1910) p. 798.

33. S. R. Thompson et al. "Evaluation of Inspection and Enforcement Programs of Other Regulatory Safety and Professional Organizations," Teknekron, Inc., Washington, DC, Vol.1, p.93.

34. Lyman, *op. cit.*

35. Thompson et al., *op. cit.*, Vol.1, p.17.

36. Leslie Maitland, "Inspection Laxities Found in Fire Department," *The New York Times*, July 13, 1980, p.1.

37. John M. Crewdson, "Violence, Often Unchecked, Pervades U.S. Border Patrol," *The New York Times*, January 14, 1980, p.D8.

38. *Ibid.*

39. Shuck, *op. cit.*, p.84.

40. *Ibid.*

41. Thompson et al., *op. cit.*, Vol.1, p.92.

42. *Ibid.*, p.89.

43. Barry M. Mitnick, *The Political Economy of Regulation: Creating, Designing and Removing Regulatory Forms* (New York, Columbia University Press, 1980), pp.238-239.

44. Barry M. Mitnick, "Compliance Reform and Strip Mining Regulation," U.S. Regulatory Council Colloquium: Innovative Regulatory Techniques in Theory and Practice, May 21, 1981, p.26.

45. "Because businessmen, lawmakers, and judges come from similar social backgrounds, are of similar age, have often been educated at the same universities, associate with the same people, and have similar outlooks on the world, it is not surprising that legislators and judges are unwilling to treat business offenders harshly." John E. Conklin, *Illegal But Not Criminal: Business Crime in America* (Prentice-Hall, 1977), p.112.

46. "Railroad Inspection Probe," *The Boston Globe*, March 28, 1981, p.4.

47. A. Gosfield, *PSRO's: The Law and the Health Consumer* (Cambridge, Mass., Ballinger Publishing Co., 1975).

48. Fred Hirschfeld, "Codes, Standards and Certificate of Authorization Program," *Mechanical Engineering*, January 1979, p.29.

49. Thompson et al., *op. cit.*, Vol.1, pp.22-25; Vol.2, pp.115-126.

50. Shuck, *op. cit.*

51. Thompson et al., *op. cit.*, Vol.1, p.ix.

52. General Accounting Office, "Supplemental Information on Assessment of the National Grain Inspection System," 1976, pp.4,31-32.

53. *Idem*, "Placing Resident Inspectors at Nuclear Powerplant Sites: Is It Working?" November 1979, p.4.

54. *Ibid.*, p.5.

55. *Ibid.*, pp.4-5.

Thompson et al., *op. cit.*, Vol.1, p.22.

56. *Ibid.*, Vol.2, p.123.

57. Danaceau, *op. cit.*, pp.35-36.

58. *Ibid.*, p.36.

59. Thompson et al., *op. cit.*, Vol.2, p.106.

60. Selwyn Raab, "Bribes to Restaurant Inspectors Alleged," *The New York Times*, December 7, 1977, p.B2.

61. "Air pollution control agencies must control a large number of sources with a limited number of personnel. This means that agencies and sources interact on an infrequent, short-duration basis, even for major sources." EPA, "Profile of Nine State and Local Air Pollution Agencies," February 1981, p.42.

62. Danaceau, *op. cit.*, p. 76.

63. President's Interagency Task Force on Workplace Safety and Health, *Making Prevention Pay* (Washington, 1978).

64. To refuse to accept a gift is "the equivalent of a declaration of war; it is a refusal of friendship and intercourse." Marcel Mauss, *The Gift* (New York, W. Norton and Company, 1967) p. 11.

65. Shuck, *op. cit.*, pp.82-83,87.

66. *Ibid.*, p.82.

67. Selwyn Raab, "Payoffs to U.S. Meat Inspectors are Found Common in City Area," *The New York Times*, April 5, 1976, pp.1,60.

68. William Robbins, "U.S. Agents Push a Broad Inquiry into Grain Trade," *The New York Times*, May 20, 1975, pp.1,25.

69. Selwyn Raab, "Bribes to Restaurant Inspectors Alleged," *The New York Times*, December 7, 1977, p.B2.

70. Thompson et al., *op. cit.*, Vol.2, p.8. Mitnick, *op. cit.*, p. 25.

71. General Accounting Office, "Supplemental Information on Assessment of the National Grain Inspection System," Attachment II, pp.4-6.

72. General Accounting Office, "Grain Inspection and Weighing Systems in the Interior of the United States—An Evaluation" April 14, 1980, p.10.

73. Paul Lehrman, "Sticker Business: So What Did You Inspect?" *The Boston Pheonix*, April 15, 1980, pp.3,12.

74. Shuck, *op. cit.*, p. 84.

75. "Another Nuclear Scandal," *Newsweek*, May 26, 1980, pp.76-78. "More Nuclear Woes," *Newsweek*, October 15, 1979, p. 87.

76. Robbins, *op. cit.*

77. Charles Radin, "Audit of MBTA Sale Brings Death Threats," *The Boston Globe*, May 21, 1980, pp.1,36.

78. "Comes for Sticker, Loses Windshield," *The Worcester Telegram*, May 15, 1981, p.1.

79. *Knapp Commission Report on Police Corruption* (New York, Braziller, Inc. 1972), pp.71-90.

80. Ed Cony, "Builders & Boodle: Bribes for Inspectors Add to Building Costs," *The Wall Street Journal*, May 16, 1961, p.1.

81. Steiger, *op. cit.*

82. Cony, *op. cit.*, p.12.

83. David Shipler, "City Construction Grafters Face Few Legal Penalties—Some Say Code Bars Jeopardizing of Jobs," *The New York Times*, June 26, 1972, p.82.

84. Cony, *op. cit.*, p.1.

85. SRI International, *Corruption in Land Use and Building Regulation.* Investigation initiated by NYC Mayor's Department of Investigation. Law Enforcement Assistance Administration, U.S. Department of Justice, September 1979, p.52.

86. Mike Tharp, "Inspection Scandals Widen, Adding to Woe of a Battered Industry," *The Wall Street Journal*, November 7, 1975, p.23.

Robbins, *op. cit.*

87. John M. Crewdson, "U.S. Immigration Service Hampered by Corruption," *The New York Times*, January 13, 1980, p.1.

88. *Idem*, "Violence, Often Unchecked, Pervades U.S. Border Patrol," *The New York Times*, January 14, 1980, p.1.

Chapter IV

Case Studies of Inspection

Nine inspection processes are examined in this chapter. They are inspections of income tax returns, inspections for occupational safety and health and inspections of housing, restaurants, grain, motor vehicles, buildings, nursing homes and elevators. While books could be written (but have not been) about each one, our purpose is succinctly to provide a greater understanding of the inspector's work environment, and to further illustrate the points made in the preceding chapters.

There are many ways to categorize regulatory inspections: for example, some inspections are federally authorized while others are mandated by state and local governments; some inspections are performed by government personnel, others by private inspectors; most inspections concern health and safety, but others do not; some regulations are directed at many industries, others at only one; sometimes the entire regulated universe is examined, sometimes only a sample; many inspections are Pass/Fail, but others use grades; most inspections involve face-to-face contact, but others do not; some inspectors work in teams, many others work alone, and so forth. The nine inspection processes described in this chapter were selected to provide sufficient variety to illustrate all of these inspection types. More importantly, however, they were chosen because they highlight many of the key environmental factors which influence the ability and the incentives of the inspector to do a good job.

I. PROBLEMS WITH ABILITY

This section describes the work of Internal Revenue Service auditors and Occupational Safety and Health compliance officers. These inspectors generally have the incentive to do a good job, but in both cases they have insufficient ability. IRS and OSHA inspection forces are so small in comparison to the regulated universe that they cannot examine more than a tiny percentage of potential inspectees. Moreover, when there is an inspection, many of the violations will excape detection. OSHA compliance officers, for example, examine behavior, which can change rapidly. Additionally, the individual inspector may visit so many diverse workplaces that he can never be knowledgeable about them all. While the IRS auditor does develop tax expertise, he has so little time per case that he can only focus on a few areas of the return. Unreported income and questionable deductions can easily slip through.

Internal Revenue Service Audits

The Internal Revenue Service (IRS) is divided into five divisions: (1) Resources Management; (2) Taxpayer Service and Returns Processing; (3) Examination; (4) Criminal Investigation; and (5) Collection. We are chiefly interested in the Examination Division where the primary IRS inspectors work. These examiners are the lowest level of the IRS hierarchy, the "street-level bureaucrats" who have face-to-face adversarial encounters with well over a million Americans each year. The Criminal Investigation Division becomes involved in a case only if the initial examiner suspects fraud and refers the case to them. The Criminal Investigation Division then decides which cases require further investigation; it is their function to develop cases beyond the reasonable doubt needed for conviction. The Collection Division is responsible for ensuring that the money owed the government is actually paid. The problem of nonfilers is a Collection activity.[1]

There are two types of IRS examiners, the revenue agent and the tax auditor. The revenue agent has an accounting background and specialized training and he handles the more complex tax returns such as those of business partnerships and corporations. Most individual tax returns are examined by tax auditors, who are often women. Revenue agents typically work in the "field" where the taxpayers' records are kept. The auditor, the focus of this analysis, stays in the office and the taxpayer brings, or occasionally mails in his records.[2]

The Auditor's Incentives

The IRS tax auditor should have the incentive to try to perform capably. The work is typically satisfying, performance can be reasonably evaluated, and the problems of capture or extortion are minimal.

The auditor has a white-collar job and works normal business hours. Job security is high, and the work can be interesting. The examiner certainly gets to meet a wide variety of people. Additionally, the auditor possesses both power and discretion. He can and does make substantive judgments about income and deductions. He is authorized, for example, to evaluate oral evidence and give reasonable allowance for certain deductions, even if written records are absent. Finally, the auditor believes in the goals of the IRS. He often displays a missionary zeal. He sees the agency and himself as acting in the public interest.[3]

The work, of course, is not without its drawbacks. Like most inspections, it does involve face-to-face adversarial confrontations, and with all types of individuals. Many auditors feel that their work, while of interest to most people, is not always appreciated by them. A lot of examiners are hesitant to reveal their occupation. Yet most are both satisfied and happy in their work.[4]

Tax auditors, unlike many inspectors, work in the office rather than in the field. Their performance is thus more easily monitored. They examine paper, which can be re-examined. Taxpayers, moreover, can request a different auditor if they are treated disrespectfully. More important, taxpayers can and often do appeal if they feel they were treated unfairly. It is not considered good performance for an auditor to have too many appeals. Not surprisingly, while some taxpayers find the auditors "highhanded and obnoxious,"[5] the large majority feel they are treated both fairly and courteously.[6]

The auditor, however, will rarely be generous with the taxpayer, for the explicitly stated goal of the IRS is to maximize yield. Indeed, some observers claim there is a quota system and those auditors with high "production" are suitably rewarded. Whether or not even an implicit quota system exists, the auditor is under pressure to increase tax revenues. If his yield is markedly below the average, he will hear about it.[7]

There is little likelihood that the IRS auditor will be captured. The auditor pays all *his* taxes, for his return is scrupulously examined, so he has little empathy for any taxpayer who may be out to cheat the government.[8] The auditor does not choose which returns to audit, nor does he know the taxpayer (he is required to disqualify himself if he does). He will typically see the taxpayer only once, for about an hour and a half. There is no continuing relationship.[9] While there is money involved, it is not big money, at least not in office audits where liabilities rarely run into the thousands of dollars. The risk of being caught accepting gratuities is thus much greater than the possible reward.[10]

Taxpayers sometimes threaten IRS auditors, but such threats are caused by loss of temper, not power tactics, and they are rarely successful. The job of the Internal Revenue Service auditor has been likened to "riding an emotional roller coaster" and "working with nitroglycerin."[11] The "combination of one auditor, one taxpayer, and one audit report can be highly inflammable, and otherwise calm, rational persons can suddenly find themselves ranting and raving."[12] Such threats die rather quickly, as the taxpayer, after all, need not accept the auditor's findings, and many do use the appeals process. The auditor, moreover, is rarely in any real danger, for a crowded government office is not a conducive atmosphere for physical abuse. Although most auditors have unlisted phone numbers, their job is a comparatively safe one. Incidents of course do occur. Recently, for example, an IRS employee was shot by an irate neighbor who believed that the agent had turned him in for non-payment of taxes.[13]

The Auditors' Ability

While each individual auditor generally has the incentive to do a good job, collectively they are unable to do so, at least in an aggregate sense. The IRS has neither the manpower nor the information to detect most tax cheating. Although all returns are superficially screened, only about 2% are actually audited. And the tax auditors cannot even do a very thorough job on these few. The auditors are so pressed for time that, typically, only selected portions of the returns are effectively scrutinized. Time pressure has created the auditors' ethic of "get the fast buck, not the last buck."[14] Effective deterence against small scale tax cheating is also lacking since the usual punishment is a requirement to pay the correct tax, plus a small dollar penalty.

The degree of tax compliance depends in part upon the ease with which transactions can be monitored. For example, the year the IRS began receiving information on interest income, the number of returns reporting such revenue jumped 45%.[15] Currently, the task of the IRS is made more difficult by the fact that it does not receive automatic or easily verifiable reports on a great deal of earnings (e.g., self-employed income; tips; capital gains; rent; etc.). Partly because of this, well over $100 billion in income is unreported every year.[16]

Certain taxes are more difficult to evade than others. One virtue of the proposed Value Added Tax (VAT), and one reason European nations adopted it, is that correct amounts are more readily collected. Not only are transactions larger and more easily monitored under VAT than for the individual income tax, but private forces help achieve tax compliance. Every buyer wants to ensure that its suppliers pay their fair amount.[17]

In current circumstances, the individual auditor has certain advantages and certain disadvantages in his encounters with the taxpayer. And, as one examiner put it, "Never forget that an audit is a form of power struggle, with both sides jockeying for advantage."[18] The principal advantage possessed by the auditor is that, except for fraud cases, most of the burden of proof falls on the taxpayer. Additionally, the auditor is generally more knowledgeable than the taxpayer about the law. On the other hand, taxpayers possess more intimate knowledge about their own income and deductions. They may also realize that the auditor has severe time constraints, and is under pressure to get taxpayer agreement. Beginning with the auditor, and in every step of the appeals process, the IRS is likely to make concessions in order to close the case.[19]

Unlike most inspections, the major burden of proof in a tax audit falls on the inspectee. This is particularly true for deductions. From the IRS perspective, it is the taxpayer who wants the deductions, so he needs to keep the receipts to prove them. When estimations are involved, the problem of proof is similarly on the taxpayer to demonstrate that his estimates are better than the auditor's. "Anytime the taxpayer guesses, the IRS has a potential adjustment."[20]

Since the auditors care about yield, they focus on those areas where revenue/minute is high. For example, since interest and tax deductions can usually be verified, they have "low audit potential."[21] Conversely, those areas where taxpayers have difficulty keeping records are ripe for scrutiny. The auditor is not so interested in whether or not it is probable that the taxpayer incurred an allowable expense, but whether he can prove it.

One reason that the burden of proof falls on the taxpayer is that, at least at the audit stage, the IRS is not judging guilt or innocence. Instead it is verifying whether the taxpayer has the receipts to match the listed expenses on the return.[22] Of course, while auditors need not accept the taxpayer's unverified word, they are permitted to evaluate oral evidence and give reasonable allowance. A primary motive for being reasonable is to achieve taxpayer's acceptance of the auditor's report in order to close the case.

Auditors possess a great deal more knowledge about tax laws and procedures than does the typical taxpayer. The auditor's position is made stronger by the fact that many citizens lack the personal confidence to challenge "official" findings.[23] Commercial preparers provide low income taxpayers with little real help in the power struggle. Employees of the major firms are usually neither attorneys nor accountants. Moreover, the IRS keeps many preparers conservative by maintaining lists of "Questionable Practitioners." If, for example, too many of the preparer's returns require audit, the IRS may begin auditing all his clients. Such a prospect is so unappealing that many preparers become little more than low-risk recorders of information.[24]

Given the complexities of the tax code, however, it often makes sense for *high* income taxpayers to hire professionals. "A taxpayer who goes to an audit interview with a lawyer specializing in tax cases, or with an accountant aware of tax court decisions, has firepower far superior to that of the typical auditor."[25] Therefore, in terms of cost-effectiveness, it is often better for the IRS to audit the small, ignorant, vulnerable taxpayer, rather that the richer, better protected one.

Interestingly, many auditors would prefer to deal with a professional preparer than the untrained taxpayer. While there is less of a knowledge advantage, there are also fewer emotional outbursts, less need for explanations, and more expeditious handling of documentation. In other words, less time is wasted. There are many things the auditor need not check. Issues usually involve interpretations of law, not facts.[26]

There is thus a second reason why individuals who file without assistance, especially those with complicated returns, are more likely to be audited. That is because they are more apt to have made grievous "errors" which need correction. Since being audited is almost always costly to taxpayers, to the extent they understand this aspect of the selection process, it should increase their incentive to use professional preparers.

The time pressure under which the auditor works significantly decreases his power. The examiner needs to close cases quickly. Although he is instructed not to horsetrade,[27] bargaining is an essential part of his work. It is implicit in his discretion to sometimes accept oral evidence, and he can more easily gain taxpayer acceptance if he sweetens the pot—"you accept this and I will not look at (or I'll accept) that."

Like many other inspectors, the auditor's decisions are influenced by his perception of the taxpayer. There is now a large amount of advice from former IRS examiners on the kinds of personal traits that may favorably dispose the auditor. These include being prompt, not overdressing, appearing frugal, acting sincere, and most important, recognizing the auditor's authority and giving him respect.[28] There is also tactical advice. The taxpayer is counseled, for example, not to announce a decision to appeal until the last possible moment. If the auditor knows an appeal is certain, he has less incentive to be "reasonable." Similarly it is best to save until the end the fact that a deductible expense was omitted from the return. Otherwise the examiner may spend the entire audit finding ways to offset the possible refund.[29]

Fraud cases usually involve unreported income rather than exaggerated deductions. The different burden of proof is undoubtedly one reason the Service tends to regard hidden income as a more serious criminal offense. While rarely used at the audit level, the IRS does have substantial investigatory powers. Special agents can gain access to the taxpayer's bank records and even his safe deposit box. They often question the taxpayer's customers, colleagues, neighbors and friends, making it apparent that the government doubts his veracity, and placing him in an embarrassing position. They can procure and pay informers. And they can legally, without a court order, use electronic surveillance.[30]

A number of factors are involved in the decision to investigate a possible fraud case. They include the amount of tax involved, the age and health of the taxpayer, and the taxpayer's prominence. Criminal prosecutions are often used for publicity purposes. Since relatively few people are audited, the Service wants a conviction to get media coverage, to increase citizens' fear of the IRS and the likelihood of vountary compliance. It even makes sense, from the IRS perspective, to audit certain celebrities. An audit of Johnny Carson, for example, that he jokes about on his show, can be worth millions.[31]

OSHA Inspection

> The amount of protection given to the laboring class is
> determined not by the number of labor laws upon the statute
> books, but by the number of such laws which are properly
> administered, and by the extent to which their provisions are
> actually enforced.[1]
>
> Bulletin of the U.S. Bureau of Labor Statistics
> February 27, 1914

The Occupational Safety and Health Administration was created in 1971 to assure, so far as possible, safe and healthful working conditions for every worker in the nation. A primary method for so assuring was to develop, adopt and enforce safety and health standards.

OSHA has been much criticized, particularly by industry. Members of the National Federation of Independent Business,

Inc. cited OSHA as their chief problem with government regulation.[2] In 1977, in a speech attacking overregulation, Jimmy Carter stated, "If I can clean up the mess in OSHA, I can be reelected."[3] And on the campaign trail, successful candidate Ronald Reagan aimed much of his invective at three enemies: Jimmy Carter, the Russian bear—and the lowly OSHA inspector.[4]

The OSHA inspector is not beloved, especially not by management. When the OSHA inspector knocks on the door, it's going to cost the company money. At minimum, it will take the time of some managers and usually some workers. In the large majority of cases, violations will be discovered. Corrections must be made and the inspector's report will often result in small fines being levied.

As a new regulatory agency, OSHA met strong initial resistance from the inspectees. Over time it has slowly become more accepted by industry. When the inspector arrives unannounced, management reaction has generally become "We realize you're doing your job, but we're sorry you're here." The attitude is less antagonistic, but the inspector's job still remains stressful.

The typical inspector wants a reasonable rapport with industry. It makes life so much easier. It also increases the likelihood of management compliance, and decreases the chance of time consuming court cases. Like many other inspectors, the OSHA compliance officer is supposed to cite all violations. But since there are usually many possible infractions—some minor, like the notorious "split toilet seat"—he will use his discretion and good judgment.[5] He will bargain, at least implicitly. The inspector may not cite an occasional infraction, or may lump violations together, in order to elicit management cooperation.

The inspector's discretion, and desire to reduce stress, raises the problem of capture. But a principal cause of the stress is also a main force preventing capture. The OSHA inspector deals directly with both parties concerning crucial economic issues. As the housing inspector comes between the landlord and tenant, so the OSHA inspector can be forced into the middle of a labor-management confrontation. The mere threat of requesting an OSHA inspection has given unions a weapon to gain concessions from management.[6]

OSHA's basic mission is to protect workers. Unlike the restaurant or grain inspector, the OSHA inspector spends time with the people the regulations are designed to help. The OSHA inspector does deal primarily with company representatives, for management controls the workplace. The firm is responsible for meeting the legal requirements and it is the firm which may be penalized for infractions. But OSHA is mandated to meet with labor as well as management. And labor can not only monitor the inspector's behavior, but offers a quite different and clearly legitimate perspective, reducing the likelihood of capture. A question arises as to who should represent labor in nonunion shops. Certainly a worker appointed by, or dependent upon management, may not provide the information and perspective often helpful for the proper enforcement of the law.[7]

There is a second important reason why the OSHA inspector will rarely be captured. Unlike meat and nuclear resident inspectors, the OSHA compliance officer visits a large number of plants. Unlike the restaurant inspector or the nursing home surveyor, the OSHA inspector also monitors a wide variety of workplaces. He has very limited contact with any particular firm, and he is exposed to many different industries. This decreases the possibility that he will become overly sympathetic to the regulated companies, or susceptible to their pressures. Certainly no single firm, or even industry, monopolizes the information flow to, or about the inspector, and none is assured of having an important influence on his subsequent career.

Capture, of course, is not impossible. Other factors, such as the background and training of inspectors affect the likelihood of capture. In its early days, most OSHA inspectors were safety specialists drawn from industry. Many undoubtedly had a managerial perspective, making them more prone to capture. But the environment of the OSHA compliance officer—examining many different industries and dealing with labor as well as management—is much less conducive to capture than many other inspection processes.

The environment of the OSHA inspector, however, does not tend to produce ideal incentives. One problem is that the work can become frustrating. The source of some frustration was the initial agency adoption of many standards which were unimportant and inappropriate. Another factor leading to frustration has been the growing bureaucratization and legalism of the agency.

As industries are said to have "life cycles," so too may regulatory agencies go through an aging process. In the early years, agencies tend to have a missionary zeal. There is more internal enthusiasm and inspectors often have more discretion. There is also usually resistance from the newly regulated industry. Mistakes are made and problems are created that need to be ironed out. As the agency ages, it learns how to avoid major confrontations. Its actions become more routine and it protects itself by becoming more rigid and bureaucratic. This aging process may have affected OSHA. Its goals have become more circumspect with time. It now creates less controversy, engenders less enthusiasm. It has become more bureaucratic, and inspectors have less formal discretionary power. There are more rules they must follow, and more forms to be completed. A number of years ago inspectors were given what they considered a silly and time-consuming one-page form to fill out. Some refused, and instead wrote up their inspection results and recommendations on the back side of the sheet. A sympathetic superior was willing to accept that because "at least they had used the form."

Confrontations with OSHA through court challenges of its decisions has made the agency increasingly legalistic in its approach. "Major emphasis is placed on assuring that detected violations can be proven" (with less emphasis on assuring that all serious hazards are detected).[8] As one inspector explained, writing a violation is like preparing a court case. "Everything must be documented."[9] The OSHA compliance officer almost always carries a camera.

It is of interest to read a couple of actual citations, one for an abrasives manufacturer, the other for a chemical corporation. They indicate the complexity and legalism in enforcement.

"29 CFR 1910.95(a): Protection against the effects of noise was not provided for employee(s) exposed to sound levels which exceeded those listed in Tables 6-16 of subpart G of 29 CFR part 1910:

The use of hearing protection was not enforced for the employee at the cupola charging station exposed to sound pressure levels 1.37 times the 8-hour permissible noise level."

"29 CFR 1910.1007(c)(3): Open vessel system operations as defined in paragraph (b) (13) of this section were not prohibited:

On the third-floor mezzanine in the AZO strike area, the glove boxes which were designed to prohibit the release into the surrounding nonregulated work areas of the regulated carcinogen 3,3' dichlorobenzidine dihydrochloride were ineffective. The leadman and the strike operator were exposed to 3.6 and 3.0 micrograms per cubic meter of air of 3,3' dichlorobenzidine dihydrochloride during the 143-minute sampling period which encompassed the dumping operation. In addition, at the sign-in desk located approximately 35 feet from the glove boxes in use, a high volume sample of the air revealed 12 micrograms of the substance."[10]

At the top of the Citation and Notification of Penalty form is the statement: "The violations described in this citation are alleged to have occurred on or about the day the inspection was made unless otherwise indicated." Good inspectors complain that the workplace is not static but dynamic, yet the legalistic approach tends to freeze the issue to the situation at the time of the site visit. Moreover, inspectors feel pushed to determine whether or not there are violations of standards rather than whether or not the workplace is hazardous.

There are two principal types of OSHA inspectors, those for safety and those for health. Safety inspections are a more traditional, accepted and straightforward part of labor regulations. Health inspection is a newer field, and in the absence of tried-and-true methods, is presently more of an art. Health inspections certainly take more time. Health inspectors require more latitude and decision-making authority to perform their job well. They are thus generally more concerned with the bureaucratization which restricts their autonomy. These industrial hygienists believe they have expertise which cannot be incorporated into Pass/Fail standards. They argue for better training of inspectors rather than ever increasing administrative control over their discretion.

The legalism and bureaucratic nature of OSHA makes it more difficult for the inspector to act as consultant and so increase his self-image. Whether or not the inspector should advise management on methods to meet the health and safety requirements is not an easy issue. OSHA has at times discouraged advice, and at other times mandated it, particularly when general rather than specific violations have been cited. OSHA does not want to give inspectors the power to bind the agency to particular corrections which it may later believe to be unacceptable. On the other hand, it seems unfair to penalize firms for general conditions without some advice as to how the situation can be rectified. At any rate, whatever the agency directives, OSHA inspectors often give informal suggestions, especially about inexpensive methods of improving workplace health and safety. And some inspectors still see their most important function in terms of education—getting management to learn a new perspective and to alter its ways.

The typical OSHA compliance officer may not have perfect incentives, but they are probably better than the incentives of many other government inspectors. On the other hand, compared to most other inspectors, the OSHA inspection force has less ability to accomplish its stated mission.

In the aggregate, the problem is that there are so many workplaces, and relatively so few compliance officers. As of 1978, for example, OSHA and state hygienists had inspected less than one percent of the Nation's estimated five million workplaces. And many of these inspections did not cover all workplace health hazards.[11]

Since the workplace is dynamic, with conditions continually changing, even a thorough inspection will not reveal many of the infractions. It is estimated, for example, that at least half of all workplace injuries involve the sort of momentary violations the inspector will be unable to detect.[12] Additionally, the inspectee can sometimes effectively hide violations. Although OSHA inspections are generally unannounced, the opening conference may give management sufficient time to alter conditions before the inspector can either see or measure them. This may be more true for health as opposed to safety hazards. Machines may be shut off, for example, decreasing pollutant and decibel levels. Of course, the presence on the inspection tour of a knowledgeable union representative can decrease the effectiveness of such tactics.

The OSHA inspector also has less legal authority than most other inspectors. While the customs inspector can probe body cavities without even "probable cause," the OSHA inspector can be denied access to a workplace until he goes through the time consuming process of obtaining a search warrant. This can eliminate the surprise aspect of many inspections.

Finally, although there is some specialization, the OSHA compliance officer typically examines so many kinds of production processes that he has difficulty being an expert on all of them. He certainly has less knowledge about what he is inspecting than an elevator, restaurant, nursing home or meat inspector. He is thus more likely to miss important violations and to cite insignificant ones.

In conclusion, we cannot expect OSHA to ensure safe and healthy working conditions for all workers. Even if all OSHA regulations were met, there would still be hazards and accidents. And, of course, all standards are not met. OSHA will never have the resources to inspect more than a tiny percentage of all workplaces in any given year. And for those which are inspected, the OSHA compliance officer will not be able to detect all, or even most of the violations.

II. PROBLEMS WITH INCENTIVES

This section examines housing, restaurant, grain, motor vehicle and building inspections. In these cases, it is not the ability of the inspector that is the main problem; it is his incentives.

Housing inspectors, particularly in slum neighborhoods, have low work satisfaction. They cannot fulfill the needs of tenants—they cannot significantly improve housing conditions—and they do not respect either the tenants or the landlords whom they most frequently encounter. Although they are not captured, they are also not that interested in doing a good job.

Restaurant, grain, motor vehicle and building inspectors are more subject to the moral hazards. Restaurant, grain and motor vehicle inspectors are likely to be captured. Restaurant inspectors meet frequently with the same proprietors and become sympathetic with their problems. In the 1970s, the individual export-grain inspectors usually worked for private firms. Many were effectively resident inspectors, monitoring a single elevator. They were easy prey to psychic inducements as well as to more tangible rewards and penalties. Their interests became closely identified with the welfare of the grain companies.

In many states, motor vehicle inspections are also performed by private companies. The motorist is not assigned to a particular inspector, but can choose from among many. Competition among inspection stations leads to superficial examinations and the overlooking of violations.

Building inspectors have the power to extort. By withholding their approval of new construction, they can cause significant and costly delays. In many cities, paying off the building inspector is the accepted practice, and probably is the only way to open a business on time.

Housing Inspection

Housing inspection can be frustrating work. One problem is that the laws are not strongly enforced. The court system is a notoriously slow method for compelling compliance with regulations. And in housing, penalties are rare, so there is little deterrence. Judges do not view housing code violations as serious crimes. Many cases are settled by a promise from the landlord that the major deficiencies will be corrected.

In 1967, the Massachusetts Housing Authority brought 400 cases to the attention of the court. No jail sentences resulted. In only eight instances were fines levied. Only one of these was paid. While the institution of the housing court has helped, overall enforcement is still slow and weak.[1]

It is not clear that strong enforcement is desirable. The codes are such that an inspector can always find some violations if he wants. While some standards are too lax and eyesores may not be infractions, all agree that literal enforcement of the housing code would prove disastrous. As one supervisor stated: "It's really a problem when you get an inspector who insists on the code."[2]

The housing authority is constrained from becoming too "successful" in bringing dwellings up to standard by the overarching economic forces that shape housing patterns. A comprehensive area inspection program can raise the quality of a deteriorated neighborhood. But this will cause rents to rise, forcing out the poor. The more common approach to housing inspection is simply to respond to complaints. This can be inequitable since many tenants in substandard units do not complain. Moreover, because a property's value will not increase by as much as the cost of improvements, piecemeal attempts to enforce the housing code may only hasten abandonments in slum areas. Not surprisingly, most complaints come from neighborhoods that already have low family income and high vacancy rates.[3]

The housing inspector cannot fulfill the expectations of the low income tenant. Therefore, he is not well liked. Indeed he is often vilified. A careful study on inspection in Baltimore concluded that housing inspection not only was "unable to generate widespread enthusiasm . . . but managed to arouse the ire of all associated with it."[4]

The work of the inspector, especially in slum neighborhoods, is filled with threats and hazards. Some carry firearms since it is dangerous for a lone official to visit a high crime area. The inspector may actually be attacked by the tenants he is supposedly assisting. During the first two months of the Baltimore inspection study, three inspectors were confronted by clients armed with pistols. Although in one case the tenant was aiming at the landlord accompanying the inspector, the inspector explained that "just the same, I ran like hell."[5]

Part of the hostility generated against slum area inspections is the lower class clients' frustrations in dealing with a middle-class dominated bureaucracy. The problem is exacerbated when the inspector is a different race than the residents of the segregated neighborhood. Most white inspectors thus prefer to be assigned to middle and upper class white neighborhoods. One inspector working in a black slum complained:

> Now I'm in another colored ghetto. I wish I could get one of those really easy assignments. I'm a nice guy. I do my inspections. I drink as much as the rest of them. So why don't I get a good job?[6]

Another inspector breathed a sigh of relief when taken off his slum area assignment:

I'm glad someone else has that area, especially in the summer-
time. Because there are too many of them (blacks) sitting outside
in the summer. You never know when one of them is going to
throw a rock or something at you. It's real dangerous.[7]

The inspector typically works 9:00–5:00, Monday through
Friday. The first hour of each morning is spent in the office,
drinking coffee, taking phone calls, and getting the day's as-
signments. At ten o'clock, the inspector begins his field visits.
He may make five or six inspections and reinspections in a day.
In the winter, lack-of-heat emergency complaints can increase
inspections to over 15 per day. Inspectors can usually ensure
that tenants receive heat. This proves to be one of the few
sources of job satisfaction.[8]

Out on assignment, a big city inspector may have difficulty
finding a parking space. When he visits a unit in a large complex, he may park a block or two away so the car does not betray
his presence. Otherwise a group of tenants will soon be clam-
oring for an inspection of their apartments. Some tenants are
not at home when he arrives, or may refuse to answer the door.
And if the tenant is present, there can be language difficulties.[9]

The inspector may be asked to cite violations that he cannot
immediately verify. The tenant, for example, may rightfully
complain of rats in the building but the inspector may see no rats,
and no evidence of current infestation. Or the tenant may claim
that the roof leaks. The inspector will have to save all his "leak-
ing-roof" complaints for a rainy day, when they can be verified.
The inspector may also be asked to cite alleged deficiencies that
in his judgment do not qualify as health hazards (e.g., a hairline
crack in a plaster wall). In 1973 in Boston, only about half of tenant complaints resulted in the inspector's finding a "cause for action," the beginning of the legal enforcement procedure.

The inspector usually views tenants with neither sympathy
nor empathy. Some tenants complain in order to avoid paying
rent. They can withhold payment while alleged violations are
under investigation, and occasionally have skipped town still
owing the last month's rent.

The most common complaints are about rubbish and trash.
The inspector arrives and finds the halls filled with debris. But

then he discovers that the interior of the tenant's apartment resembles a pigsty. He thinks the tenants should be more self-reliant and accept more responsibility for the condition of the apartment house. He may wonder "Who threw the garbage in the hall?" And the answer is, "Not the landlord—who lives 250 miles away." The inspector may begin to refer to tenants, especially disliked ethnic groups, as "animals." While infrequent, he can even cite the tenants for violations of the good-house-keeping provisions.[10]

The inspector also does not have much love for the landlord. Most face-to-face encounters occur because the landlord is providing substandard housing. The owners that take up most of the inspector's time are the worst slumlords. Those who "milk" their property often meet with the inspector's "official wrath."[11] One inspector remarked to us that some landlords and tenants "deserve each other."

The inspector has discretion about what, if anything, to cite. A major influence on his decision is his view of the worthiness of the particular tenant and landlord. Resident landlords who make sincere attempts to carry out repairs merit more consideration than do recalcitrant absentee landlords. Tenants whose complaints are substantial and who evince high standards of housekeeping are similarly considered more deserving.

There are few problems of extortion by housing inspectors. While they have some discretion, housing inspectors have little real power. Capture is also unlikely. The housing inspector is caught in a dispute between tenant and landlord. He hears both sides. He may be sympathetic with neither. He sees large numbers of tenants and owners and rarely becomes friendly with any of them. Landlords occasionally try to bribe or intimidate the inspector, but, in general, since enforcement is weak, why bother? While housing inspectors may accomplish little, they are rarely on-the-take.

Overall, the housing inspector has the ability to do a good reporting job. It is not difficult to detect most major deficiencies. The landlord cannot hide many problems from both the tenant and the inspector. In the aggregate, of course, if inspectors only respond to complaints, they may not have the opportunity to examine all substandard dwellings. Additionally, the need to rely on their own observations can create minor difficulties. But the main problem is that while inspectors can do a good reporting job, they usually do not have the ability to significantly improve housing conditions.

Inspectors often point to the lack of court enforcement as the major culprit behind their ineffectiveness. But the more fundamental cause is that the agency's mission is suspect. It is not clear that major, forced improvements in slum housing conditions are desirable when they mean higher rents and increased abandonments.

The inspector's inability to help causes resentment on the part of the residents, and frustration on the part of the inspector. Slum tenants come to dislike and abuse the inspector. The inspector, in turn, learns to regard low income tenants with contempt. Many problems, he feels, are caused by their own slovenly behavior. One inspector told us he was tired of their "endless complaining."

The slum inspector's work brings little satisfaction. Job quality and job usefulness are low. While there is not much danger of capture, the inspector has little incentive to do a good job. Instead, his incentive is often to do as little as possible.

Restaurant Inspection

In the opening scene of *Invasion of the Body Snatchers*, Donald Sutherland, an honest and aggressive inspector, discovers some evidence of rat infestation in a French restaurant. He cites the problem, much to the annoyance of the proprietor. When he leaves, he finds the windshield of his automobile has been shattered—presumably by the restaurant employees. It would have been safer for Sutherland to have failed to report the problem or he might have tried to extort payment to keep quiet about it. Instead, he did his job and fell victim to verbal and physical harassment.

Physical intimidation is the exception rather than the rule in actual restaurant inspection. But it can happen. The adversarial nature of the inspector-restaurateur relationship and their face-to-face encounters provides an environment where intimidation can occur.

Restaurants are public places. Customers form personal opinions on ambience, decor, service and cuisine. However, when it comes to hygiene and sanitation, the customer is at the mercy of the proprietor. Customers assume that the food is prepared correctly in a clean kitchen. If they are wrong, they may never know it. Or they may not know which meal—and hence which restaurant—is the culprit. There is thus a need to ensure that restaurants provide healthful conditions in the storage, handling and preparation of food.

Restaurants are required to obtain annual operating permits from the local health department which enforces the food laws. As a result, they must be inspected each year. More frequent inspections, however, are typical, and these are essential if the system is to work successfully. The inspections are usually unannounced and routine, although some may be initiated by complaints. Complaints often are from disgruntled employees and prove groundless.

The restaurant inspector examines both equipment and behavior. An equipment problem might be a walk-in refrigerator that is not operating at a safe temperature. Or a leaky ceiling could threaten the food stored below. Behavioral problems include dirty can-openers, unwashed hands, open garbage cans, unswept floors, excess grease in the exhaust vents, and long unnetted hair. These behavioral problems can be quickly corrected if the inspection is anticipated. They might easily recur the day after the inspection. By having frequent unannounced inspections, and repeatedly flagging violations, the inspector may gradually be able to force compliance. Although little dollar outlay is required to maintain sanitary conditions, many owners would otherwise let conduct worsen, saving the time and trouble of policing for cleanliness.

For very serious problems, such as severe cockroach infestation, the inspector may shut down the restaurant on the spot. In San Francisco, for example, a magazine food editor accompanying an inspector witnessed the legal closing of a restaurant within two minutes after the discovery that the running water had been shut off due to a sewerage backup.[1] Additionally, if food is spoiled, the inspector may immediately dump Clorox or motor oil on it to ensure that it will never be served.[2] For most problems, however, the enforcement process is slow and drawn out.

When the inspector leaves, he gives the manager a written report that lists all violations. It also specifies the date on which a "compliance" reinspection is scheduled, commonly seven days after the original inspection. If the noted conditions have not been corrected by that time, a hearing day will be scheduled. The manager has another chance, for the inspector will double-check the restaurant the day of the hearing to make sure that legal action is necessary. After the hearing, a notice of closure can be issued. There is yet another opportunity for correction since the closure notice does not become effective for one to seven days. Only if the problems have still not been remedied will the restaurant actually be shut down.

Closure is a harsh penalty and is imposed infrequently. Some communities, however, have no other formal sanction. To some extent, then, citations for a few minor deficiencies can be safely ignored. But the inspector can impose costs on chronic offenders by having frequent reinspections. Still, only after the threat of a shutdown becomes real will some owners agree to correct infractions.

The restaurant inspector is often a "sanitary inspector" who also inspects retail food stores, bakeries and swimming pools. Both the restaurant and housing inspectors enforce the prevailing Sanitary Code, concentrating on different provisions. They often share the same office. In smaller municipalities, the same person may hold both positions. Typically, he prefers his job as restaurant inspector. The work is much more rewarding.

Many housing inspectors are frustrated. Vigorous enforcement of the housing code could economically injure the same tenants the law was intended to protect. The inspector's efforts are thus rarely supported by effective enforcement. Most of his field time is spent with tenants he has little respect for, and whose expectations he cannot fulfill. By contrast, restaurant regulations are not controversial. The inspector is backed by stronger enforcement. He can and does improve sanitary conditions without causing undue hardship and without creating additional problems.

In many communities, the restaurant inspector's job is reasonably well paying and fairly easy. An inspector might do four or five inspections in a day; many can be done in less than a half-hour. The workday of these inspectors consists of periods of productive work interspersed with longer intervals of seemingly less productive time.

The restaurant inspector's job is sometimes considered a political plum. Inspectors are naturally sensitive to the notion that they may receive favored hiring treatment or that they do less than a full day's work. One inspector told us the "three unwritten rules" for the health inspector. First, take care of personal business early in the day—there may not be time in the afternoon while out in the field. Second, never leave your car in front of your home for more than forty-five minutes during lunch. If you're going to take longer, park the car at least a block or two away. Finally, never drink on the job—even if free drinks are offered.

As these rules imply, while the inspector does not work in obscurity, there is little effective oversight. As is true for most other street-level bureaucrats, especially those working in the field, it is difficult to evaluate the performance of the restaurant inspector. In large cities with sizable inspection departments, there may be "senior inspectors" with administrative responsibilities who accompany line inspectors and rate their performance. Or a recent report may be pulled at random and verified by another inspection of the restaurant. Unfortunately, there are serious problems with both methods of evaluation. The inspector may behave differently while being observed. And many restaurant conditions—especially behavioral problems—can change overnight.

While the restaurant inspector is generally satisfied with his work, he is prone to capture. He interacts continually and exclusively with the restaurateur. While he may have some empathy with customers, he does not meet them on his job. Almost all complaints about his performance come from restaurant owners and managers.

For the typical restaurant inspector, the regulated universe is small. He is likely to know personally and be known by the restaurateur. Indeed, he may spend part of his working day chatting over coffee with various managers. Restaurant owners have also been known to use monetary bribes as well as gentle persuasion to win over the inspector. It has been reported, for example, that payoffs by Manhattan restaurateurs to health inspectors are "both widespread and routine."[3]

The danger of capture can sometimes be reduced by rotating inspectors among different zones. A problem occurs, however, if this makes the inspector-inspectee relationship more impersonal, for then it could also make the inspector less effective. Since most infractions are not of sufficient magnitude to close a restaurant, the inspector must rely on the good faith and voluntary compliance of management. An inspector who gains a reputation as a cold personality or as a stickler who cites every violation may only succeed in annoying and angering the proprietor. The restaurateur who feels the inspector is "out to get" him may not cooperate. The coffee break with the proprietor may be viewed not as unproductive time, but as an investment in winning the restaurateur to the inspector's side, increasing the degree of compliance with the regulations. Seen from this perspective, while Donald Sutherland may not have been captured, he probably was not the most effective inspector.

There is not only the danger that restaurant inspectors may be captured, but also the possibility of extortion by them.[4] The inspector may be in a position to demand tribute from an owner by threatening to close the restaurant. But since shutdowns are rare, the threat may not be entirely credible. The inspector can also harass the proprietor with frequent inspections at which numerous minor violations are cited. Of course, owners of unsanitary restaurants should feel harassed, for an accepted method of coercing recalcitrant restaurateurs into compliance is by repeated inspections. Complaints by owners have sometimes forced inspectors to travel in pairs to defend themselves against accusations of extortion.[5] Yet, to the extent that there are moral hazards, the problem is usually due to capture by the restaurateur rather than to extortion by the inspector.

Grain Inspection

For many years, serious problems existed in our national grain inspection system. Bribery and corruption were not uncommon. The situation came to a head in the mid-1970s when

57 indictments were handed down against grain companies and inspection firms.[1]

The grain that is inspected by U.S. authorities includes corn, wheat, rye, oats, barley, flaxseed, sorghum, soybeans, mixed grain and feed grains. Grain inspection can involve grading, weighing, and examining storage facilities. There is mandatory inspection for grain that is exported and voluntary inspection for grain shipped internally within the United States.

While it has not been without problems, the internal system has functioned more effectively. This is principally because the market is somewhat self-policing. Buyers can identify the supplier of the grain as well as the inspection agency. "Anyone dissatisfied with grades assigned in a marketing area can always refuse to base future purchases on grades assigned in that market, or they can buy grain from someone else next time."[2]

Purchasers of U.S. exports have more difficulty pinpointing the exact source of underweight or misgraded grain. A poor product casts a bad light on all U.S. shipments rather than on any particular inspection agency or supplier. That is why it is important for our nation to have a reputable inspection system. It is also why inspection problems are more likely to occur in the export market.

The grain scandals of the 1970s were export scandals. As a result, the Grain Standards Act of 1976[3] was passed, changing the U. S. export inspection system. The voluntary system for the inspection of internal grain was left virtually intact. This discussion focuses on the deficiencies of pre-1976 export grain inspection.

The fundamental problem with the system was not in the ability of the inspectors to do a good job. It was with their incentives. The actual grain inspectors worked for a number of federally licensed private firms, trade associations and state agencies. There were sometimes conflicts-of-interest. Grain companies and their officers could own and control the licensed private firms that officially inspected their product. For example, the majority of members on the board of directors of one inspection agency were officials in the grain companies served by that agency. The board appointed the agency's general manager as well as its chief inspector. In addition, the grain companies owned stock in the inspection agency.[4] That was literal capture.

There were also the more traditional forms of capture. Working alongside elevator employees, the inspectors tended "to develop relationships and attitudes favorable to elevator interests" and to become prey for special gratuities—"liquor, meals, tickets to sporting events and office parties," and even bribes.[5]

The likelihood of capture was especially great for the resident inspectors who monitored the same firms for long periods of time. For some of the small inspection agencies, the possibility of rotating personnel was severely limited. Of the 26 designated agencies inspecting export shipments, 17 made inspections at only one or two elevators. A number of licensed inspectors in the New Orleans area had worked at a single elevator for over 15 years.[6]

The independence of the inspectors was further compromised by the fact that the agencies often received free space from the grain companies for their private laboratories.[7]

Added to all this was the shipper's ability, at least in some instances, to choose the agency that would do the inspection. Sometimes samples were submitted from the same lot to a number of agencies. This practice, commonly called "shopping for grade," placed the inspectors in a competitive position.[8] This kind of competition is not usually beneficial for society, especially given that the private inspectors depend on the inspectees for their income. The reward system favors those who grade leniently. This was, and continues to be, more of a problem for interior grain shipments.

Not only is export grain inspected, but the vessel stowage area is also examined to ensure that it is not contaminated. While stowage inspections can be made quickly, costly shipping delays can result if violations are cited that require correction. In this situation, the grain inspector has a power similar to the building inspector whose approval must be given before occupancy is permitted. Indeed, it was an inspector's excessive extortionary demand that led to a ship captain's complaint and ultimately resulted in the public exposure of the widespread corruption that existed in the export grain inspection system.[9]

The United States Department of Agriculture (USDA) had

difficulty monitoring the behavior of the privately licensed inspection agencies. Federal supervisors reported verbal and even physical harassment when they visited the port elevators. They were jostled, their car tires were slashed, and a contract was even put out on one supervisor who had cited violations.[10] There were also the more mundane problems of monitoring. The heterogeneous nature of most grain lots, for example, means a resample will not exactly duplicate the original. A few resamples may thus not be sufficient to determine if the inspector was following prescribed procedures.

The USDA also made it hard on itself. When original samples were regraded, the licensed inspector was sometimes permitted to choose the particular ones to be re-examined. He thus tried to select those he believed to be free of error. Additionally, the value of federal spot checks of grain inspection *procedures* was limited by the fact that the licensed personnel, and the elevator management, were usually aware that they were being observed. Since the behavior of inspector and inspectee can be rapidly and easily modified, unobtrusive monitoring is generally advantageous. Yet in some places USDA gave notice of their visits; at others, the federal supervisors wore bright orange overalls and were thus easily recognized.[11]

The Grain Standards Act of 1976 seems to have removed the most blatant problems in export inspection. Conflicts of interest between grain merchandisers and inspection agencies have been largely eliminated. Private agencies are no longer authorized to perform these inspections, and state agencies have been investigated before receiving licenses. New federal supervisory teams were also created to more thoroughly monitor export inspection activities. Yet, as a recent GAO report documents, export grain inspection, similar to most other inspection systems, is still not without its deficiencies.[12]

Motor Vehicle Inspection

There is a variety of approaches to motor vehicle safety inspection in the United States. We will contrast two representative, but quite different approaches—the one employed by the District of Columbia, the other by the state of Massachusetts.

In the District of Columbia, there are a small number of

government-run stations, staffed by full-time inspectors. These inspectors are civil service employees, with a minimum of five years' experience as mechanics. Their task is not only to thoroughly examine every vehicle licensed in the District, but also to help educate car owners about safety, and to help discover stolen vehicles.[1] Many repair shops have located close to these few inspection stations.

By contrast, in states such as Massachusetts, Pennsylvania and Virginia, thousands of private stations are licensed to inspect automobiles. Inspection is performed by private individuals whose principal occupation is working in gas stations or repair shops. Inspection comprises only a small part of a typical working day.

An interesting aspect of motor vehicle inspection in these states, as well as in the District of Columbia, is that the motorist goes to the inspector. A more critical factor is that the motorist can choose among inspection stations. The large number of geographically dispersed inspection stations in states like Massachusetts decreases the time and inconvenience cost to the car owner. However, this also reduces the effective monitoring of inspector behavior. The large number of stations eliminates the extortion power of the inspector. Unfortunately it also greatly diminishes his incentive to do a good job.

Inspectors in all these places have the ability to perform capably. Anyone can determine whether the horn, turn signals and windshield wipers work. Some specialized knowledge and equipment is required to check other safety features such as brake linings. But generally the task is relatively simple and straightforward. A motorist cannot hide deficiencies from a competent inspector.

In the aggregate, the entire inspection force can ensure high vehicular safety standards, at least in the short run. Although there are a large number of motorists, in most inspection states all cars are examined. This is feasible since the actual inspection requires only a few minutes. After receiving his sticker, the owner may let his automobile deteriorate, but he has no incentive actively to attempt to make his car less safe.

The crucial issue is not whether motor vehicle inspectors have the ability, but whether they have the incentive to do a good job. In Massachusetts, Pennsylvania and Virginia, they do not.

One problem is the difficulty of monitoring and rewarding inspector performance. Because there are so many geographically dispersed stations, the state authorities have a tough time overseeing daily activity. Massachusetts, for example, has 34 state inspectors whose job it is to monitor over 3,000 motor vehicle inspection stations.[2]

Unlike elevator inspections, inspector performance cannot be evaluated sometime after-the-fact. The motor vehicle inspector has merely affirmed that, at a specific time, minimum safety conditions were met. He cannot be held responsible for the car's condition two days later. Liability is thus not an issue for motor vehicle inspectors. This is all the more true since it is difficult, if not impossible, to pinpoint the cause of most accidents. As a result, insurance companies have never relied on motor vehicle inspections to certify safety.

Not only is monitoring Massachusetts inspectors difficult, but so also is the rewarding of good and the punishing of poor performance. The inspectors do not work for the state; they are private individuals working for private firms. The state cannot give specific inspectors pay hikes or promotions, nor can it fire incompetents. All it can do is either designate or deny firms as official inspection stations. In the health field, this is called "institutional" as contrasted to "occupational" licensure. The difference in this situation is that inspection is not usually a significant part of the institution's total activities. Most stations become inspection sites primarily as a small service for their customers.

The more severe incentive problem for Massachusetts motor vehicle inspectors is capture. While there is some reason to be strict—since the inspection station may be asked to do the required repair work—most inspectors appear to be far too lenient.

Inspection requires face-to-face contact, often with people known to the inspector. It is not pleasant to fail a car, since repairs may be costly for the motorist, and re-inspection an added hassle. Additionally, while being strict might lead to complaints, a lenient inspection almost never does.

The key cause of capture, however, is not such considerations but the fact that the motorist can choose among numerous inspection stations. Individually, motorists generally prefer a fast, Passing inspection, and this is what they get. The market works. There is a kind of Gresham's Law operating, with lenient inspectors driving out the strict ones. Many stations do a quick, superficial job, maximizing inspection revenue/minute and attracting more motorists. Others feel required to pass the questionable vehicles of regular or potential customers in order to maintain or garner their goodwill.[3]

Economists are typically interested in the optimal. If we are locked into the Massachusetts inspection approach, then the optimal number of inspection sites may be fairly low. Large numbers of stations increase competition. And competition promotes lenient inspections. These are desired by the individual motorist, but are not necessarily in the interest of the public at large.

There is general consensus that automobile safety inspections are not done well.[4] Now auto emissions are being subjected to inspection. The emissions will typically be checked by the same people who inspect for safety, often at the same time. We should not, therefore, expect too much from this emission monitoring. In states such as Massachusetts, the inspectors will have the same poor incentives. In all areas they will be less able to ensure effective compliance with the law. The principal reason has to do with the incentive of motorists. The automobile owner has little reason to try to decrease his car's safety. But in order to increase gas mileage, he may want to tamper with the emission controls. The day after inspection, many vehicles deliberately may be in violation of the new law. A good inspection in this case does not ensure good performance, even in the very short run.

Building Inspection

Almost every structure in the United States is subject to a multitude of mandatory building code regulations. The public official immediately responsible for insuring that these requirements are met is the building inspector.

The inspector is typically an older man with a building background. New York City, for example, requires that each inspector have at least nine years' experience in the building trades.[1] The inspector's job is not highly prestigious. The salary is modest. The inspector is often the lowest-paid man on a construction site.[2]

The building inspector has a number of duties. Most familiar is the task of monitoring ongoing construction for conformance to the building code, and issuing a Certificate of Occupancy to a newly completed building. But the inspector also responds to complaints from residential tenants concerned with the structural soundness of their dwellings. He conducts the annual safety inspections of places of assembly, which include churches, schools, meeting halls, and taverns. Nursing homes and hospitals undergo annual inspections. In smaller towns, building inspectors may also examine for elevator safety.

There is a major difference in the inspection environment between overseeing new building construction and examining buildings already in use. In the former, the inspector's approval is required before occupancy is permitted. The inspector thus has great power, which sometimes is misused. For existing structures the balance of power is reversed. Occupancy continues until the Building Department demonstrates unsafe conditions. In the real world this means the inspector often finds it difficult to insure the quick correction of safety problems.

New Construction

Building inspectors make the news principally when scandals expose widespread corruption. New York is the major city—but certainly not the only one—in which payoffs and bribery have proved to be chronic problems. In 1966, a Grand Jury report stated:

> The basic facts of the current pattern of corruption in the Plan Section of the Department of Buildings are engagingly simple: if you want prompt service and fair treatment you have to pay for it. If you do not pay you may be subjected to interminable delays, "lost" files, highly technical objections, or other harassment. It is as simple as that: pay or else.
>
> Who gets paid? Almost everyone: Clerks, Plan Examiners, Multiple Dwelling Examiners and Plumbing Inspectors.[3]

In 1974, an undercover investigator posed as a building in-spector. He was under specific instructions to do nothing to solicit bribes, yet he was offered payments in 44 of 66 instances. The total amount of money paid to him was over $2,500.[4]

The crucial factor giving the building inspector power is that his approval is required before work can continue or occupancy is permitted. Inspector procrastination can thereby impose huge costs on the contractor. If the inspector fails to appear on sched-ule, expensive delays may result. Paying the inspector to show up on time is thus not unusual.[5]

A second element in the corruption problem is the hundreds of pages of detailed building standards that are codified into law by most local or state authorities. Many of the standards are excessively restrictive; others are out-of-date. Even the most exacting architect and builder will break some of these stan-dards. Indeed, state-of-the-art workmanship may be out of compliance. The code book thus gives the inspector a legal justification for delaying any new project.[6]

Despite such lawful authority, some builders complain that inspectors have threatened to discover imaginary faults. A plumber alleged that one inspector found leaks where they did not exist. The inspector could have required extra work to "fix" them had he not been paid off.[7]

Small variations in the inspector's performance can mean thousands of dollars to the building owner. Payoffs are a way to buy time, to decrease construction costs, and to reduce bu-reaucratic hassles. Inspectors are just one of several groups in a favorable position to receive payments from contractors. Po-lice, city clerks, highway officials, FHA agents and even union personnel have been on the take. Elected officials have also been known to put pressure on the inspector "not to make waves" and to "lay off" a particular development.[8]

It is not in the interest of the individual builder to fight a corrupt system. Great *private* costs to the builder will accompany any attempt to provide a *public* good—restoring honest inspec-tion. One lawyer who tape-recorded extortion attempts, and whose work resulted in the discharge and arrest of a New York City inspector, encountered great difficulties at the Building Department when she subsequently represented a developer. A few years later she found herself unable to secure a Certificate of Occupancy for the supermarket she wished to open.[9] It is little wonder that few people are willing to fight City Hall, and that bribes become a normal cost of doing business.

Once corruption becomes commonplace, it also becomes suicidal for the individual inspector to buck the system. One will not make friends by revealing the malfeasance of co-workers and superiors. And every inspector who has taken bribes risks penalty once the corruption is exposed. Additionally, potential future income from bribes is substantially decreased. A New York City inspector in the 1970s could more than double his income by accepting bribes.[10]

In some localities, seemingly none of the inspectors is on the take. And in all areas, many inspectors are honest as well as capable. But the inspection environment in new construction allows and often promotes corrupt behavior. This is particularly true for major construction projects, especially during building booms. Then the inspector's time is limited, and any delay will cause large dollar losses.[11]

Existing Structures

For many inspectors, directly supervising new construction is only a small component of their job. A large part consists of doing paperwork and occasionally making court appearances. Most of the time actually spent in the field is consumed by annual safety inspections or in responding to tenant complaints. As with complaints to the Housing Department, these are sometimes designed simply to forestall rent increases. Such cases exhaust and dispirit the inspector.

It is much harder to shut down a deficient structure than to prevent a new one from opening. Time is on the side of the owner. Due process usually works against strict code enforcement. The inspector often feels unsupported by the courts. Penalties are rare, and so is deterrence. Judges generally believe that finally forcing a recalcitrant owner into code compliance is a satisfactory solution.

It is much harder for a corrupt inspector to extort tribute from the owner of an existing building than from one with facilities under construction. Procrastination by the inspector imposes few, if any, costs. To extract payoffs in these instances generally requires clear harassment—not impossible, but certainly a more questionable, difficult, and conspicuous procedure. Not surprisingly, almost all the building inspection scandals have involved new construction.

Conclusion

A force of trained building inspectors has the ability to do a good job, at least for new construction. This does not mean that the inspectors will see to it that every minor, possibly outdated standard in the building code is met. It means, instead, that the inspectors can ensure that all new buildings are safe. First, every new structure can be inspected. There is no sampling. Secondly, it is a building rather than behavior that is being examined. The contractor has great difficulty hiding a serious deficiency from a competent inspector. Additionally, the inspector has the legal power to ensure that all violations are corrected. Finally, because there is little reason for the contractor to deliberately make a building less safe after it has been approved, a good inspection will achieve long lasting compliance with the law.

The building inspectors have the ability. The problem is with their incentives. In many areas, inspector corruption is endemic. Sometimes there is capture-the inspector permitting the contractor to build to less than the desired level of quality or safety, usually in exchange for hard cash. More often, however, corruption takes the form of that rarer of moral hazards—extortion. In many cities, greasing the inspector's palm has become a normal business expense, even for the highest quality builders.

III. BETTER REPORTING

The final two case studies deal with nursing home and elevator inspections. Originally, this section was entitled "More Successful Inspections," which might imply that the regulatory system itself was successful. But that does not seem to be the case for nursing homes.

Nursing home surveyors can't turn the facilities into wonderful places to live. Often, the improvements induced by the regulations are fairly marginal. But most nursing home surveyors seem to have the motivation, along with some ability, to do a good job in the reporting of violations. While surveyors will, of course, miss many behavioral infractions, they should be able to discover the major problems at each and every facility.

Of all the inspections analyzed, elevator examinations should be the most "successful." Elevator inspectors have the incentive and the ability to detect all important violations at all elevators. And in this case, a good inspection should mean good elevator performance. The elevator inspector alone has the power to pretty much ensure passenger safety.

Nursing Home Inspection

Nursing home inspections are called surveys. They can be long and detailed. A housing or a restaurant inspection could be done in an hour or so by one person. A *team* of inspectors might spend a week at a nursing home. A thorough certification inspection involves an examination of almost all of the important aspects of the nursing facility. In Massachusetts, surveyors are required to rate each of 627 items on a scale from 0 (excellent) to 5 (terrible).

The surveyors inspect the medical, dietetic, rehabilitation, pharmaceutical, laboratory, dental and social services provided for the patients. They examine the physical environment, infection-control procedures and disaster preparedness. They scrutinize the medical record, nursing plans, and medical-care-evaluation records. All patient activities are under their purview.

The surveyors spend most of their time reading records. But they also make direct observations of the facility, noting understaffing or overcrowding, poor housekeeping, or the improper following of written procedures. They watch the meals being prepared and served, the medications being prepared and administered, the patients engaging in physical and social activities. They also receive some information from limited discussions with the doctors, patients, nurses and other staff. In general, the surveyors try to ensure that there is more than just paper compliance with the regulations.

Outcome measures are examined, as well as input variables. Surveyors will examine medical records to determine if there have been improvements in the physical health of patients. The positioning of patients is noted, as is the response to conversations. Lack of alertness, or especially a display of fear, may indicate that the home is not providing the right kind of care. The surveyor uses such outcome observations to get a feel for the general quality of the nursing home. If there are problems, they must be documented on the more input-focused rating sheet.

Some of the citations a high-quality nursing home in Massachusetts received in 1980 included:

1. A lack of documentation that every employee had received the required TB test;
2. A lack of documentation that the medical director had reviewed all medical activities;
3. A lack of documentation that ambulation was provided for some patients who requested it;
4. A lack of documentation that restraints were provided for some patients who required them;
5. Some medical progress reports were not sufficiently updated;
6. A lack of sufficient assessment of patients' interests in activity plans;
7. Certain closets that should have been locked were not;
8. Emergency drug kits were not easily available.

As can be seen, the law—and surveyors—require a large amount of paper documentation.

Citations—minor citations—in 1981 included somewhat inappropriate dietary plans for some patients, a dirty kitchen fan, the placement of a dishrack on the floor, orderlies transporting laundry without using bags, and some patients not seeing a doctor for more than sixty days.

The nursing home administrator expects some citations. It is an extremely rare event when none is given. The provider at a good home often rationalizes the citations with a "We try to run a good home, but nobody's perfect." There is a tendency to blame others— the patient's particular doctor, or employees- for many of the problems. And the workers can be a problem. In most facilities, there are more employees than there are patients.

Administrators are somewhat concerned about the results of the nursing home survey. For one thing, it is always more personally satisfying to receive higher rather than lower marks. For another, the results are made public, which may influence the home's ability to attract the more lucrative private patients. Additionally, there is, for some, the possibility of decertification or even delicensure. Finally, and potentially the most important, has been the attempt, at least in Massachusetts, to tie reimbursement rates to survey results. Fascinatingly, while the Massachusetts nursing homes are told how the reimbursement system weighs the 627 items, the surveyors themselves are not supposed to know. It is an opposite situation from the IRS where the inspectors have reasonably accurate information about agency priorities, and try to keep the public in the dark.

It is primarily the publicity and the incentive reimbursement system that help make nursing home surveys more than just an exercise in attempting to upgrade the very worst homes. If inspection were simply Pass-Fail, with failure leading to possible decertification and closure, there would not be great pressure on average or good homes to improve their care. In the current situation of excess demand for nursing home care, and given the fact that the physical moving of patients from one facility to another is both emotionally traumatic and medically dangerous, the threat of decertification and closure is not very credible. Tying reimbursement to survey results creates some marginal financial incentives for all facilities.

The inspectors themselves would also emphasize that they help to improve good homes by providing advice and motivation. Their job is to sell health care. They see themselves partly in a role of consultant and teacher, providing a slightly different, yet expert perspective, and keeping the administration from becoming too complacent. They believe they help the administration as well as the patients. They have a saying: "Good facilities deserve good surveyors. Bad ones need them."

While surveyors could theoretically perform a useful function as consultants, they are not always effective in this role. In the first place, they are not trained as either teachers or consultants. Their professional experience makes them better at looking at details rather than abstracting from these to larger management problems. Additionally, as happened with OSHA inspectors, a growing legalism in the system has decreased the possibility for a beneficial exchange of ideas. The attempt to close the facilities of the worst repeat offenders in Massachusetts caused surveyors increasingly to be used as gatherers of detailed, legally admissible evidence and less as assessors of the general quality of care. It is indicative that in the exit interview, surveyors are now required to read verbatim the specific citations that will appear in their final report.

Nursing home surveyors are more likely than many other inspectors to have the incentive to try to do a good job. Quality regulations—and thus inspections—make more sense for nursing home care than for many other services. Neither the exit nor the voice option work well in providing market incentives rewarding high quality. Patients are not always rational, nor good judges of many aspects of quality. Moreover, complaints are often counterproductive for the individual who is relatively powerless. For patients already settled in a home, the costs of exit may be too great to make it a viable option. Exit is also constrained by the excess demand created by governmental policies. The one area where market forces might theoretically be effective—in the initial selection of a home—is also severely limited by the existence of excess demand.

Nursing home surveyors thus see themselves as performing a necessary and important role in society. They cite with pride what they see as a major upgrading of care since the surveys began (e.g., the dramatic improvements in state schools for the mentally retarded). They grow concerned—for the welfare of patients—if they become excluded from monitoring the care provided to private as distinct from public patients.

In Massachusetts, the surveyors must be Registered Nurses with a certain amount of experience. This means the surveyors are almost always women, usually middle-aged. Some intimidation (harassment) of these women, while not usual, is certainly not unheard of. They may be yelled at, or asked to leave the facility. In bad neighborhoods they may be told they had better continually check their car to insure against vandalism. However, the fact that they often operate as a team helps to protect them against verbal and physical pressures or abuse.

Since the surveyors deal continually and exclusively with nursing homes, there is the distinct possibility of "capture." As is the case for many inspectors, there is a tendency to try to turn a potentially adversarial situation into one of cooperation.

It is particularly difficult to always be ruthlessly honest in the exit interviews when surveyors must give face-to-face evaluations to the administrator. Not only is there an incentive to minimize job-related conflict situations, but surveyors may become familiar with and sympathetic to the administrator's problems. This may be especially true since surveys can run up to a week or more. Additionally, a tough report rarely brings any tangible benefit to surveyors—either in terms of pay, promotion or prestige.

Surveyor decisions can be challenged—principally by the facility. The surveyors must therefore substantiate their judgments (not by photographic documentation or laboratory tests as is often the case for OSHA inspectors, but by specific written example). There is also some direct bargaining with the provider, even about the wording of the reports. At one home, for example, an issue arose over whether an area should be termed "dirty" or merely "soiled." The challenges by and confrontations with the providers mean that surveyors may impose citations somewhat reluctantly, again raising the spectre of capture.

But there are many other, usually stronger forces which decrease the likelihood of capture. Massachusetts surveyors are R.N.'s. They are professionals. Their primary career identification has been as nurses. They have professional training and support groups. When they work in teams this support is immediate. Nursing home administrators, moreover, do not control their future careers. There are a large number of other areas where R.N. services are both needed and in demand.

The training of nurses emphasizes patient care. As surveyors, they believe that good patient care and the quality of life remain as their major objectives. Moreover, they see and talk to patients. Their time is *not* spent exclusively with providers. Finally, and more importantly, they not only sympathize but emphathize with the patients. They talk continuously of their relatives going to or being in a nursing home; they also believe that some day they too may become nursing home patients.

Compared to most inspectors, nursing home surveyors not only have the incentive, but they also have the ability to do a good reporting job. The surveyors usually work as a team, and help each other. They are not reluctant to call on specialists—physical therapists, social workers, etc.—who may also be inspectors. And because the surveyors are able to scrutinize the home so intensely, they often discover violations of other laws, such as labor violations, Medicaid fraud, etc.

Nursing home surveys are usually unannounced, however, since certification requires inspection once a year, most facilities know approximately when the surveyors will be coming. This general knowledge, however, should not significantly affect the inspection report. The surveyors can examine the medical records, payrolls, time sheets, patient activities, administration minutes, etc. for the entire period. Additionally, current behavior is carefully monitored. While this could theoretically be improved specifically for the survey period, given the facts: (a) that inspectors can talk to patients; (b) that there are large numbers of employees, many sympathetic to the surveyors; (c) that there are many outside personnel (e.g., doctors) involved in the care; and (d) that the survey takes a long time, it should be very difficult to completely hoodwink a competent surveyor. It should be noted, however, that even the best surveyors will probably never catch the isolated night nurse who fabricates a few records.

Since our field trips with the nursing home surveyors, a new method of inspection has been instituted in Massachusetts. Facilities are now categorized as "outstanding", "acceptable" or "marginal" by the most recent inspection results. The better homes receive a shorter, more superficial audit, with ratings for only 50 items. This strategy originally had been designed to leave more survey time available to intensively investigate the worst facilities. However, major appropriation cuts in the summer of 1981 reduced the number of surveyors available to inspect Massachusetts's 550 nursing homes from 59 to 25. This meant that no homes have actually received additional scrutiny under the new "survey by exemption."[1]

Elevator Inspection

Elevators have been called the "safest form of mechanical transportation." Malfunctions usually cause inconvenience, but they rarely result in injury. Accidents, of course, do happen. In the 18-month period ending January 1981, there were 29 fatalities associated with elevators in the United States. The large majority of deaths occurred when people somehow got into the elevator shaft. Some stepped into the shaft when the doors opened but the car was not there. A mechanic fell off the top of an elevator and was crushed by the car. One elderly woman was killed when she was knocked over by a closing elevator door. No fatalities resulted from elevators' free-falling down the shaft. A type of fail-safe system virtually eliminates this particular problem. The most frequent cause of injury is when the car does not stop level with the floor and people trip and fall.

Even without mandatory inspection, elevators would be reasonably safe. Preventative maintenance is generally the best strategy. Elevators have long lifetimes—the majority of elevators in the United States were built before 1930. Thus repair expenditures can be amortized over many years. It is also very costly to replace an old elevator, and there is not really a second-hand market for used elevators. Additionally, there is liability. The building owner is ultimately accountable for elevator problems. And the cause of any elevator accident is almost always traceable to mechanical defects, making it possible to pinpoint responsibility.

Owners typically contract for regular maintenance service. This entails such things as lubricating the cables, oiling and greasing the motor and pulley bearings, dusting the shaft (cutting a fire hazard) and examining and cleaning the electrical switches. There are some prepaid complete-service contracts that provide motor replacement, cable replacement, and all other repairs, but these are quite expensive. Most contracts simply call for periodic surveys of elevator condition.

Some owners have more incentive to maintain their elevators than do others. A department store manager is well advised to keep the elevator safe and functioning smoothly. So too is the owner of a prestigious center city office building. Less inclined to purchase high-priced, high-quality maintenance contracts are less conspicuous owners with captive users. Elevators in a housing project or a rent-controlled apartment building may be poorly serviced. The owner of a building which is soon to be demolished may also refrain from spending money on elevator maintenance. It is in such circumstances that inspection can significantly improve elevator operation and safety.

Every state has a legal requirement that elevators be kept in safe working order. However, the states have different rules for safety inspection. In New Hampshire and Rhode Island, for example, the safety inspection is performed by private elevator mechanics, who then report to a state inspector, who independently verifies some of their findings. In Massachusetts, a government inspector must be present while the mechanics conduct the annual safety inspection. A striking aspect of this type of inspection is the collaboration between the inspector and private mechanics.

The quality of elevator-repair firms varies. An inspector may trust the work of a particular company, but with others he may be more cautious. Rather than simply observing mechanics performing the tests, the inspector may actually participate, especially in checking the cables, where visual inspection alone may not be adequate.

Inspecting a six- or eight-story elevator takes about 30 to 45 minutes. There should, of course, be no set time limit because of the problems that may arise. In Massachusetts, the inspector makes an appointment to meet at the elevator with a team of mechanics hired by the owner. These are generally from the maintenance service company. The mechanics check the motor and the condition of the pulleys. Bearings are examined; they may be lubed or marked for replacement. Then "into the hatch": the inspector and mechanic get on top of the car, ride up and down and check the door stops, counterweights and cables. Mechanics look for small breaks in the wires that make up the strands, that in turn make up the cables. Too many breaks indicate the cable is worn and should be replaced. Cable safety factors are based on cable speed. A common requirement is for each one of the four or more cables to have sufficient strength to hold *five times* the weight of the car at its maximum design load. Cables generally last from five to fifteen years.

After inspection of the shaft, the "safeties" are checked. These brakes automatically activate should the car begin to move at a greater than preset speed. Last, the "pit" below the bottom floor is examined. Here the mechanic may find money. On some safety inspections the elevator is loaded to capacity with ballast and tested to make sure it handles the weight.

If an elevator has a poor performance record, it may be inspected more than once a year. But typically the safety-related items deteriorate rather slowly, and any nascent problem that may manifest itself within the year can be detected during the inspection.

Small maintenance defects that turn up during the inspection are often corrected on the spot. The presence of repairmen simplifies the repairs. Adjustments are made to mechanical switches, or uneven shaft guide rails are filed smooth. Larger maintenance problems are left for another time. If they are serious, they are ordered to be corrected as early as the following day.

Elevator inspectors generally have the ability to do a good job. Their experience and training are usually sufficient. In Massachusetts, for example, a newly appointed elevator inspector must be a state-licensed elevator mechanic. This entails at least two years' apprenticeship as a mechanic's helper. However, a grandfather clause exempts existing inspectors from this requirement, so not all may have this much training.

Inspectors usually examine every elevator, rather than merely taking a representative sampling. They also have the time to perform thorough examinations. Additionally, owners typically have neither the knowledge, nor the desire, to hide deficiencies. Although the regulatory penalty structure does not provide for marginal deterrence, most owners will quickly order all required repairs.

The inspector has the ability to do a good job, and a good inspection assures good performance. Not only is the elevator tested under actual operating conditions, but equipment problems are virtually the sole source of failure. The inspector also has enforcement power. He can condemn an unsafe elevator on the spot, and allow it to reopen only after it has been repaired to his satisfaction.

About 60 percent of elevators receive their permits at the time of inspection. Most of the remainder get theirs within 60 days, after the required repairs and reinspection. Very few are "posted"—taken out of service because an imminent danger exists. The inspector must weigh the consequences of condemning an elevator. In a building of elderly tenants, for example, closing down the only elevator may cause serious problems. An inspector may sometimes risk keeping such an elevator open until repairs can be made.

Elevator inspectors also generally have the incentive to perform well. There is little danger of capture. The building owner cannot choose inspectors, but must accept the one assigned. Moreover, unlike most inspections, there is no face-to-face encounter with the inspectee. Usually the inspector has no contact with the owner whatsoever. (One inspector told us that he tried to remain ignorant of repair costs to prevent financial considerations from affecting his judgment.) In those instances when the building owner does meet the inspector, his ignorance of elevator operation puts him at a distinct disadvantage.

The inspector will not normally meet strong resistance when he orders expensive elevator repairs. Usually the owner is neither knowledgeable nor present, and the maintenance firm is not sorry to have the added repair business. Indeed, any competent mechanic will flag items that clearly need repair. Both the reputation of his firm and his personal license are potentially at stake.

There are, however, gray areas in elevator safety, where judgment is required and experts could disagree. The inspector might then find some opposition, especially if the maintenance company's incentive is to cut costs by not requiring repairs. This could be the case if the owner has purchased a prepaid full-service contract. The individual mechanic may believe the repair is probably necessary but feels pressure to resist the added expense. The inspector's order relieves his problem of how to justify the cost to the service company management or to the owner of the building.

Overall, the elevator inspector has little reason not to perform capably. The importance of his task is not generally questioned. Shoddy work will not save him time—the inspector observes and assists at an inspection that is carried out by others. Additionally, poor work, if it results in accidents, can result in legal liability. The inspector rarely has any face-to-face contact with the building owner. And when he does, his knowledge is generally vastly superior. Indeed, what pressure he may feel often comes from the service company—which usually promotes excessive rather than deficient safety.

NOTES AND REFERENCES

Internal Revenue Service Audits

1. Report to the Joint Committee on Internal Revenue Taxation, "Audit of Individual Income Tax Returns by the Internal Revenue Service," 1976, GGD-76-55, p. 44.
Mary L. Sprouse, *How to Survive a Tax Audit* (New York, Doubleday & Co., Inc., 1981), p. 3.
2. *Ibid.*, p. 4.
3. Jeff Schnepper, *Inside IRS* (New York, Stein & Day, 1978), pp. 46-47.
Sprouse, *op. cit.*, p. 38.
Paul Strassels and Robert Wool, *All You Need to Know About the IRS* (New York, Random House, Inc., 1979), p. 5.
4. Schnepper, *op. cit.*, p. 38.
Milton Lehman, "Nobody Loves the Collector," *The Nation's Business*, November 1951, p. 33.
5. Schnepper, *op. cit.*, p. 26.
6. Report to the Joint Committee on Internal Revenue Taxation, *op. cit.*, GGD-76-54, p. 42.
7. Strassels and Wool, *op. cit.*, pp. 39-40.
Sprouse, *op. cit.*, p. 88.
Schnepper, *op. cit.*, pp. 16-17, 31-32.
8. Strassels and Wool, *op. cit.*, p. 100.
Lillian Doris (ed.), *The American Way in Taxation* (Englewood, New Jersey, Prentice-Hall, Inc., 1963).
9. Sprouse, *op. cit.*, p. 92.
10. *Ibid.*, p. 218.
11. *Ibid.*, p. 6.
12. *Ibid.*, p. 222.
13. "IRS Agent, Wife Shot," *The San Francisco Chronicle*, February 22, 1982.
14. Sprouse, *op. cit.*, p. 20.
15. Frederick Andrews, *Tax Tips and Tax Dodges* (New York, Dow Jones & Co., Inc., 1971).
16. Joseph O'Donoghue, "Everything You Never Wanted to Know About the IRS," *Commonweal*, March 28, 1980, p. 174.
17. *Ibid.*, pp. 180-181.
18. Sprouse, *op. cit.*, p. 30.
19. Strassels and Wool, *op. cit.*, p. 105.
Sprouse, *op. cit.*, p. 189.
20. *Ibid.*, p. 125.
21. *Ibid.*, p. 127.
22. *Ibid.*, p. 44.
23. Report to the Joint Commission on Internal Revenue Taxation, *op. cit.*, GGD-76-54, p. iii.
24. O'Donoghue, *op.cit.*, p. 173.

25. *Ibid.*
26. Strassels and Wool, *op. cit.*, pp. 13,82.
27. *Ibid.*, p. 142.
28. Sprouse, *op. cit.*, pp. 95-98.
Strassels and Wool, *op. cit.*, Chapter 10.
29. Sprouse, *op. cit.*, pp. 97-98.
30. George Hansen and Larrey Anderson, *How the IRS Seizes Your Dollars and How to Fight Back* (New York, Simon & Schuster, Inc., 1981), p. 13.
Schnepper, *op. cit.*, pp. 112-113, 187.
31. Sprouse, *op. cit.*, pp. 233-235.
Strassels and Wool, *op. cit.*, p. 9.

OSHA Inspection

1. George Price, "Administration of Labor Laws and Factory Inspection in Certain European Countries," Bulletin of the U.S. Bureau of Labor Statistics, Feb. 27, 1914, p. 9.
2. "What's All the Uproar over OSHA's 'Nit-Picking' Rules?" *National Journal*, 10/7/78, p. 1595.
3. *Ibid.*, p. 1594.
4. David Pauley et al., "A Deregulation Report Card," *Newsweek*, January 11, 1982, pp. 50-52.
5. John Mendeloff, *Regulating Safety: An Economic and Political Analysis of Occupational Safety and Health Policy* (Cambridge, Mass., MIT Press, 1980), p. 40.
6. *Ibid.*, pp. 10-30.
7. Matt Witt and Steve Early, "The Worker as Safety Inspector," *Working Papers*, September/October 1980, p. 28.
8. General Accounting Office, "Workplace Inspection Program Weak in Detecting and Correcting Serious Hazards," May 19, 1978, p. 9.
9. Paul Danaceau, "Making Inspection Work: Three Case Studies," United States Regulatory Council, May 1981, p. 73.
10. Citation and Notification of Penalty for two Ohio firms, 1980.
11. General Accounting Office, "Sporadic Workplace Inspections for Lethal and Other Serious Health Hazards," April 5, 1978, p. 10.
12. Mendeloff, *op. cit.*, p. 128.

Housing Inspection

1. Pietro Salvatore Nivola, "Municipal Agency: A Study of Housing Inspection in Boston," Harvard Ph.D. dissertation, May 1976, p. 140.
2. Virginia Boyle Ermer, "Street Level Bureaucrats in Baltimore: The Case of Housing Code Enforcement," The Johns Hopkins Ph.D. dissertation, 1972, p. 64.
3. Nivola, *op. cit.*, p. 25.
4. Ermer, *op. cit.*, p. 27.
5. *Ibid.*, p. 59.

6. *Ibid.*, p. 62.
7. *Ibid.*, p. 59.
8. Michael Goodwin, "Cold Means Feverish Work for City's Boiler Inspectors," *The New York Times*, February 17, 1979, p. 24.
9. Nivola, *op. cit.*, Chapter 4.
10. *Ibid.*, Chapter 5.
11. Ermer, *op. cit.*, p. 77.

Restaurant Inspection

1. Stan Sesser, "On the Prowl with the Kitchen Police," *San Francisco Magazine*, April 1982, p. 57.
2. Cheryl Morrison, "Looking for Trouble," *New York Magazine*, March 3, 1980, p. 100.
3. "Sticky Fingers, Greasy Spoons," *The New York Times*, December 18, 1977, Sect.4, p. 18, column 1.
4. Cf. Russel Pergament, " 'Shanghaied' in Brookline: Health Inspector Arraigned in Alleged Extortion Attempt," *Newton Tab*, February 6, 1980, p. 1. "City Health Inspector Charged in Extortion," *The Boston Globe*, June 12, 1982, p. 24.
5. Raab Selwyn, "Bribes to Restaurant Inspectors Alleged," *The New York Times*, December 7, 1977, p. 82.

Grain Inspection

1. William Robbins, "U.S. Agents Push a Broad Inquiry into Grain Trade," *The New York Times*, May 20, 1975, p. 25.
William Robbins, "Inquiry Widening on Grain Exports," *The New York Times*, June 25, 1975, pp. 1,18.
Mike Tharp, "Inspection Scandals Widen, Adding to Woe of a Battered Industry," *The Wall Street Journal*, November 10, 1975, p. 23.
William Robbins, "Grain Scandals Harm U.S. Sales," *The New York Times*, November 30, 1975, pp. 1,65.
2. General Accounting Office, "Grain Inspection and Weighing Systems in the Interior of the United States—An Evaluation," April 14, 1980, p. 10.
3. United States Grain Standards Act of 1976 (7.U.S.C. 71 *et seq.*)
4. General Accounting Office, "Supplemental Information on Assessment of the National Grain Inspection System," July 16, 1976, p. 13.
5. *Ibid.*, pp. 13,32.
6. *Ibid.*, p. 32.
7. *Ibid.*, Attachment II, pp.11,28.
8. *Ibid.*, Attachment II, pp.4,6.
9. Tharp, *op. cit.*, p. 23.
10. William Robbins, "U.S. Agents Push a Broad Inquiry," *The New York Times*, May 20, 1975, pp.1,25.
11. General Accounting Office, "Supplemental Information on Assessment of the National Grain Inspection System," *op. cit.*, p. 20.

12. General Accounting Office, "Federal Export Grain Inspection and Weighing Programs: Improvements Can Make Them More Effective and Less Costly," November 30, 1979.

Motor Vehicle Inspection

1. "Too Dangerous to Drive," *Popular Mechanics* 127:99-101, January 1967.
2. Conversation with Massachusetts inspection supervisor, June 1982.
3. "Sticker Business: So what did you inspect?" *The Boston Phoenix*, April 15, 1980, pp.3,12.
4. F. Baily, "State Auto Inspections: A $2 Nuisance?" *Better Homes & Gardens*, 51:48-51, April 1973.

Building Inspection

1. David Shipler, "Study Finds $25 Million Yearly in Bribes is Paid by City's Construction Industry," *The New York Times*, June 26, 1972, p.1.
2. David Shipler, "City Construction Grafters Face Few Legal Penalties," *The New York Times*, June 27, 1972, p. 82.
3. Cited in SRI International, *Corruption in Land Use and Building Regulation*, U.S. Department of Justice, September 1979, p. 53.
4. *Ibid.*, p. 52.
5. Shipler, June 27, 1972, *op. cit.*.
6. Shipler, June 26, 1972, *op. cit.*.
7. Shipler, June 27, 1972, *op. cit.*.
8. SRI International, *op. cit.*, p. 46.
9. Shipler, June 26, 1972, *op. cit.*
10. Carter Horsley, "Construction, Corruption and Delays," *The New York Times*, April 29, 1976, p. 30.
11. SRI International, *op. cit.*, p. 47.

Nursing Home Inspection

1. Kathryn Tolbert, "Inspections change for Nursing Homes," *The Boston Globe*, October 2, 1981, p. 19.

Conclusion

The broad goal of this treatise has been to increase our understanding of inspection. I hope that, having finished the book, when you hear that food is "USDA approved" or that the "building has been inspected," you are more likely to think about, and better able to understand what such assurances really mean.

The book focuses on the inspector. He is the primary agency contact with the inspectee and ofttimes with the public. Without him there is virtually no enforcement. His actions are thus critical to the successful implementation of the regulations. The inspector, however, does not have a glamorous job. His is the lowest paying and least prestigious professional position in the agency. He is at the bottom of the regulatory totem pole. He has also been generally neglected by policy analysts.

This book examines those factors that influence the inspector's ability and incentive to do a good job. Our basic thesis is that there is an important relationship between the inspection environment and inspector performance. If we understand the setting in which the inspector works, we can make useful predictions concerning how well he will do his job.

An inspector may have a number of tasks, including the reporting of violations, consulting with the regulated parties, and directly enforcing the law. Every inspector acts as a Reporter, and it is this particular function that the book emphasizes. The aspect of the inspector's job that most differentiates it from other occupations is its adversarial nature. As a reporter, the inspector is supposed to find and cite deficiencies. The inspectee prefers that few infractions be either discovered or reported.

The book distinguishes between the individual inspector's ability to do a good job, and the capacity of the entire inspection force for discovering violations. It also differentiates between the inspector's ability to do a good reporting job and his ability to help accomplish the agency's mission. The best Reporter is not necessarily the best inspector.

There are three main factors affecting the inspector's ability to do a good reporting job. The first two, which we call: (1) Innate Ability and Training; and (2) Difficulty of the Task, help determine whether a worker in any profession could capably perform his duty. The third factor, Balance of Power, focuses on the adversarial nature of inspection, the fact that the inspectee may actively attempt to conceal his deficiencies.

Many features of the inspection environment affect the inspectee's ability to hide violations. Among the most important are whether it is behavior or equipment that is being examined (it is easier to momentarily alter behavior), whether there are knowledgeable parties who will complain or inform, and the legal access accorded to the inspector.

Having the ability to do a good job is, of course, not sufficient. One must also have the incentive. Three broad factors determine whether the inspector will try to perform capably. The first two, Work Satisfaction and Performance Evaluation, are important for virtually any job. (Compared to most low level employees, field inspectors receive little scrutiny or supervision, so their performance is particularly difficult to evaluate.) The third factor, which we term the Moral Hazards, encompasses both capture and extortion, and emphasizes the conflict inherent in the inspector-inspectee relationship.

Face-to-face encounters, continual interaction, absence of third-parties—these are the types of characteristics that make the inspector prone to capture. The resident inspector, although potentially better able to do a good reporting job and to elicit inspectee cooperation, is more susceptible to capture, to cooptation. The inspector whose approval is required before the inspectee can proceed with operations is the one most capable of extorting tribute.

Many interesting and important aspects of inspection are not covered in this short book. For example, this is not a "how to" book, telling directly either how to be a good inspector or how to be a good agency administrator. However, by providing a broad perspective and background information, the book should prove helpful in better understanding both of these tasks.

One of my students suggested that the flyer for this book should read, "Have you ever paid income tax? or worked in a factory? or rented an apartment? been in a building? eaten in a restaurant? ridden in an elevator? driven a car? If so, then this book is for you." I admire the student's pluck, for inspection is ubiquitous and affects everyone. This book presents a conceptual framework for thinking about this important topic.

List of Acronyms

ASME	–	American Society of Mechanical Engineers
CPSC	–	Consumer Product Safety Commission
EPA	–	Environmental Protection Agency
ERDA	–	Energy Research and Development Administration
FAA	–	Federal Aviation Administration
FBI	–	Federal Bureau of Investigation
FDA	–	Food and Drug Administration
GAO	–	General Accounting Office
INS	–	Immigration and Naturalization Service
IRS	–	Internal Revenue Service
MBTA	–	Massachusetts Bay Transportation Authority
MESA	–	Mining Enforcement and Safety Administration
MPIP	–	Meat and Poultry Inspection Program
NASA	–	National Aeronautics and Space Administration
NRC	–	Nuclear Regulatory Commission
OSHA	–	Occupational Safety and Health Administration
PSRO	–	Professional Standards Review Organization
USDA	–	United States Department of Agriculture

Bibliography

Books

Andrews, Frederick. *Tax Tips and Tax Dodges*. New York: Dow Jones & Co., Inc., 1971.

Bardach, Eugene and Kagan, Robert A. *Going by the Book: the Problem of Regulatory Unreasonableness*. Philadelphia: Temple University Press, 1982.

Bartlett, John. *Familiar Quotations*. Boston: Little Brown, 1980.

Conklin, John E. *Illegal But Not Criminal: Business Crime in America*. New York: Prentice-Hall, Inc., 1977.

Doris, Lillian (ed.). *The American Way in Taxation*. Englewood, New Jersey: Prentice-Hall, Inc., 1963.

Feldman, Penny and Roberts, Marc. "Magic Bullets or Seven Card Stud: Understanding Health Care Regulation" in Richard S. Gordon, *Issues in Health Care Regulation*. New York: McGraw-Hill, 1980.

Galbraith, John Kenneth. *The New Industrial State*. Boston, Houghton-Mifflin Co., 1967.

Gosfield, A. *PSRO's: The Law and the Health Consumer*. Cambridge, Massachusetts: Ballinger Publishing Co., 1975.

Hansen, George and Anderson, Larrey. *How the IRS Seizes Your Dollars and How to Fight Back*. New York: Simon & Schuster, Inc., 1981.

Hemenway, David. *Industrywide Voluntary Product Standards*. Cambridge, Massachusetts: Ballinger Publishing Company, 1975.

Klinger, Donald E. *Public Personnel Management*. New York: Prentice-Hall, 1980.

Leighton, Gerald and Douglas, Loudan. *The Meat Industry and Meat Inspection*. London: Educational Book Co., 1910.

Lipsky, Michael. *Street Level Bureacracy*. New York: Russell Sage Foundation, 1980.

Mauss, Marcel. *The Gift*. New York: W. Norton and Company, 1967.

Mendeloff, John. *Regulating Safety: An Economic and Political Analysis of Occupational Safety and Health Policy.* Cambridge, Massachusetts: MIT Press, 1980.

Mitnick, Barry M. *The Poltical Economy of Regulation: Creating, Designing and Removing Regulatory Forms.* New York: Columbia University Press, 1980.

Niederhoffer, Arthur, et al. *The Ambivalent Force: Perspectives on the Police.* San Francisco: Rinehart Press, 1973.

Rose-Ackerman, Susan. *Corruption: A Study in Political Economy.* New York: Academic Press, 1978.

Schnepper, Jeff. *Inside IRS.* New York: Stein & Day, 1978.

Sprouse, Mary. *How to Survive a Tax Audit.* New York: Doubleday & Co., Inc., 1981.

Strassels, Paul and Wool, Robert. *All You Need to Know About the IRS.* New York: Random House, 1979.

Wilson, James Q. *The Investigators: Managing FBI and Narcotics Agents.* New York: Basic Books, 1978.

· · · · ·

Knapp Commission Report on Police Corruption. New York: Braziller, Inc., 1972.

President's Commission of Law Enforcement and Administration of Justice. *The Challenge of Crime in a Free Society.* New York: Avon Books, 1968.

Periodicals

Alexander, Rosemary. "Unwarranted Power at the Border: the Intrusive Body Search," *Southwestern Law Journal,* Vol. 32, November 1978.

Alexander, Tom. "Why Bureaucracy Keeps Growing," *Fortune,* May 7, 1979.

Baily, F. "State Auto Inspections: A $2 Nuisance?" *Better Homes & Gardens* 51; April, 1973.

Birnbaum, Stephen. "Customary Procedures," *Playboy,* June, 1980.

Cassedy, James. "Applied Microscopy and American Pork Diplomacy: Charles Wardell Stiles in Germany: 1898-1899," *ISIS,* Vol. 62, Spring, 1971.

Clark, Timothy, B. "What's All the Uproar Over OSHA's 'Nit-Picking' Rules?" *National Journal,* October 7, 1978.

Diver, Colin. "A Theory of Regulatory Enforcement," *Public Policy,* 28(3), Summer, 1980.

Hardy, B. "Poultry Men Feel Inspection Pinch," *Farm Journal,* Vol. 83, May, 1959.

Henkin, Louis. "Arms Inspection and the Constitution," *Bulletin of the Atomic Scientists,* Vol. XV, No. 5, May, 1957.

Hirschfeld, Fred. "Codes, Standards and Certificate of Authorization Program," *Mechanical Engineering*, January, 1979.

Kimoto, H. "More Than Just a Sticker?" *Modern Photography*, Vol. 38, April 1974.

Kirschenbaum, Jerry. "Steroids: The Growing Menace," *Sports Illustrated*, November, 1979.

Lehman, Milton. "Nobody Loves the Collector," *The Nation's Business*, November, 1951.

Lenorovitz, Jeffrey. "DC-9 Tailcone Loss Spurs Aft Bulkhead Inspections," *Aviation Week and Space Technology*, September 24, 1979.

McKean, Roland. "Enforcement Costs in Environmental and Safety Regulation," *Policy Analysis*, 6(3), Summer, 1980.

Midgley, Elizabeth. "Immigrants: Whose Huddled Masses?" *Atlantic Magazine*, April, 1978.

Morrison, Cheryl. "Looking for Trouble," *New York Magazine*, March 3, 1980.

O'Donoghue, Joseph. "Everything You Never Wanted to Know About the IRS," *Commonwealth*, March 28, 1980.

Pauley, David et al. "A Deregulation Report Card," *Newsweek*, January 11, 1982.

Rothblatt, Henry. "Income Tax Evasion: Dealing With the IRS Special Agents and Prosecutions," *Criminal Law Bulletin*, June, 1974.

Sesser, Stan. "On the Prowl with the Kitchen Police," *San Francisco Magazine*, April, 1982.

Shuck, Peter. "The Curious Case of the Indicted Meat Inspectors: Lambs to Slaughter," *Harper's*, September, 1972.

Sussman, O. "New Law's Dangerous Deceit," *The Nation's Business*, May, 1968.

Tedesco, J. "Let the Camera Be Your Eyes," *The American City*, January, 1979.

Witt, Matt and Early, Steve. "The Worker as Safety Inspector," *Working Papers*, September/October, 1980.

. . . . .

"Another Nuclear Scandal," *Newsweek*, May 26, 1980.

"Bill Vindicated – Inspectors May Need Warrants," *Time*, 111:16, June 5, 1978.

"Manufacturer to Monitor DC-10 Engine/Wing Pylon," *Aviation Week*, July 23, 1979.

"More Nuclear Woes," *Newsweek*, October 15, 1979.

"Now OSHA Must Justify Its Inspection Targets," *Business Week*, April 9, 1979.

"Too Dangerous to Drive," *Popular Mechanics*, 127; January, 1967.

Newspaper Articles

Cony, Ed. "Builders & Boodle: Bribes for Inspectors Add to Building Costs," *The Wall Street Journal*, May 16, 1961.

Crewdson, John. "U.S. Immigration Service Hampered by Corruption," *The New York Times*, January 13, 1980.

Idem. "Violence, Often Unchecked, Pervades U.S. Border Patrol," *The New York Times*, January 14, 1980.

Diehl, Jackson. "Tired of Complaining; Tenants Lose Hope," *The Washington Post*, July 23, 1979.

Goodwin, Michael. "Cold Means Feverish Work for City's Boiler Inspectors," *The New York Times*, February 17, 1969.

Henderson, Scott. "Thefts, Losses Pose Perennial Headache," *The Harvard Independent*, March 6–12, 1980.

Horsley, Carter. "Construction, Corruption and Delays," *The New York Times*, April 29, 1976.

Lehrman, Paul. "Sticker Business: So What Did You Inspect?" *The Boston Phoenix*, April 15, 1980.

Maitland, Leslie. "Inspection Laxities Found in Fire Department," *The New York Times*, July 13, 1980.

McCombs, Phil. "Emissions Testing Coming: 350,000 Cars Will Flunk," *The Washington Post*, July 8, 1981.

Montville, Leigh. "A Peek Inside of the Room TV Never Will See," *The Boston Globe*, February 13, 1980.

Newell, David. "Drugs, Fruit & Technology are Customs' Varied Quarry," *The New York Times*, July 5, 1981.

Pergament, Russel. " 'Shanghaied' in Brookline: Health Inspector Arraigned in Alleged Extortion Attempt," *The New Tab*, February 6, 1980.

Quinn, Jane Bryant. "The Income Tax Cheat May Be Finding It Easier," *The Boston Globe*, April 6, 1981.

Raab, Selwyn. "Bribes to Restaurant Inspectors Alleged," *The New York Times*, December 7, 1977.

Raab, Selwyn. "Payoffs to U.S. Meat Inspectors are Found Common in City Area," *The New York Times*, April 5, 1976.

Radin, Charles. "Audit of MBTA Sale Brings Death Threats," *The Boston Globe*, May 21, 1980.

Robbins, William. "Grain Scandals Harm U.S. Sales," *The New York Times*, November 30, 1975.

Idem. "Inquiry Widening on Grain Exports," *The New York Times*, June 25, 1975.

Idem. "U.S. Agents Push a Broad Inquiry into Grain Trade," *The New York Times*, May 20, 1975.

Shipler, David. "City Construction Grafters Face Few Legal Penalties—Some Say Code Bars Jeopardizing of Jobs," *The New York Times*, June 27, 1972.

Shipler, David. "Study Finds $25 Million Yearly in Bribes is Paid by City's Construction Industry," *The New York Times*, June 26, 1972.

Steiger, Paul E. "New York, Los Angeles Strive to Halt Payoffs to Building Inspectors," *The Wall Street Journal*, December 7, 1966.

Tharp, Mike. "Inspection Scandals Widen, Adding to Woe of a Battered Industry," *The Wall Street Journal*, November 7, 1975.

Tolbert, Kathryn. "Inspections Change for Nursing Homes," *The Boston Globe*, October 2, 1981.

Ulman, Neil. "Navigation and the Troubled Tankers," *The Wall Street Journal*, Februry 15, 1977.
Zweig, Phillip. "Technology Goes to Sea," *The New York Times*, June 3, 1977.

. . . . .

"City Health Inspector Charged in Extortion," *The Boston Globe*, June 12, 1982.
"Comes for Sticker, Loses Windshield," *The Worcester Telegram*, May 15, 1981.
"IRS Agent, Wife Shot," *The San Francisco Chronicle*, February 22, 1982.
"Railroad Inspection Probe," *The Boston Globe*, March 28, 1981.
"Sticky Fingers, Greasy Spoons," *The New York Times*, December 18, 1977.
"18 Food Places Called Health Code Violators," *The New York Times*, July 19, 1981.

Government Documents

Bittner, Egon. "Functions of the Police in Modern Society; a Review of Background Factors, Current Practices and Possible Role Models," National Institute of Mental Health, Center for Studies of Crime and Delinquency," Chevy Chase, Maryland, 1970.
Danaceau, Paul. "Making Inspection Work: Three Case Studies," U.S. Regulatory Council, May 1981.
Lyman, Theodore, Program Director. "Corruption in Land Use and Building Regulation," SRI International, U.S. Department of Justice, September, 1979.
Mitnik, Barry M. "Compliance Reform and Strip Mining Regulation," U.S. Regulatory Council Colloquium: Innovative Regulatory Techniques in Theory and Practice, May 21, 1981.
Price, George. "Administration of Labor Laws and Factory Inspection in Certain European Countries," Bulletin of the U.S. Bureau of Labor Statistics, February 27, 1914.
Thompson, S.R. et al. "Evaluation of Inspection and Enforcement Programs of Other Regulatory, Safety and Professional Organizations," for the Office of Inspection and Enforcement, U.S. Nuclear Regulatory Commission, Contract #NRC-05-77-064.
Vinz, George, L. "The Compleat Inspector," *FDA Papers*, September, 1968.

. . . . .

Egg Product Inspection Act, 91st Congress, 2nd Session, Report #91–16, December 3, 1970.
Environmental Protection Agency, "Profile of Nine State and Local Air Pollution Agencies," February, 1981.
General Accounting Office, "Audit of Individual Income Tax Returns by the

Internal Revenue Service," Report to the Joint Committee on Internal Revenue Taxation, GGD–76–54,55, 1976.

General Accounting Office, "Federal Export Grain Inspection and Weighing Programs: Improvements Can Make Them More Effective and Less Costly," November 30, 1979.

General Accounting Office, "Grain Inspection and Weighing Systems in the Interior of the United States—An Evaluation," April 14, 1980.

General Accounting Office, "Heroin Being Smuggled into New York City Successfully," Bureau of Customs, Department of the Treasury, Bureau of Narcotics—Dangerous Drugs, B–164031(2), December 7, 1972.

General Accounting Office, "Improvements Still Needed in Coal Mine Dust-Sampling Program and Penalty Assessments and Collections," December 1975.

General Accounting Office, "Placing Resident Inspectors at Nuclear Power Plant Sites: Is it Working?", November 15, 1979.

General Accounting Office, "Prospects Dim for Effectively Enforcing Immigration Laws," November 5, 1980.

General Accounting Office, "Sporadic Workplace Inspections for Lethal and Other Serious Health Hazards," April 5, 1978.

General Accounting Office, "Supplemental Information on Assessment of the National Grain Inspection System," Attachments I and II. Department of Agriculture, CED–76-132, July, 1976.

General Accounting Office, "Workplace Inspection Program Weak in Detecting and Correcting Serious Hazards," 1978.

President's Interagency Task Force on Workplace Safety and Health, "Making Prevention Pay," Washington, 1978.

United States Grain Standards Act of 1976.

Doctoral Dissertations

Broberg, Merle. "A Study of a Performance Control System as an Indicator of Organizational Goals: The Housekeeping Inspection System of the Philadelphia Housing Authority," Ph.D. Dissertation in Sociology, American University, 1969.

Ermer, Virginia Boyle, "Street Level Bureaucrats in Baltimore: The Case of Housing Code Enforcement," Ph.D. Dissertation, The Johns Hopkins University, 1972.

Nivola, Pietro. "A Municipal Agency: A Study of Housing Inspection in Boston," Ph.D. Dissertation, Harvard University, May, 1976.

Index